Mathematical Basics of Motion and Deformation in Computer Graphics

Second Edition

Synthesis Lectures on Visual Computing
Computer Graphics, Animation, Computational
Photography, and Imaging

Editor
Brian A. Barsky, *University of California, Berkeley*

This series presents lectures on research and development in visual computing for an audience of professional developers, researchers, and advanced students. Topics of interest include computational photography, animation, visualization, special effects, game design, image techniques, computational geometry, modeling, rendering, and others of interest to the visual computing system developer or researcher.

Wang Tiles in Computer Graphics
Ares Lagae
2009

Virtual Crowds: Methods, Simulation, and Control
Nuria Pelechano, Jan M. Allbeck, and Norman I. Badler
2008

Interactive Shape Design
Marie-Paule Cani, Takeo Igarashi, and Geoff Wyvill
2008

Real-Time Massive Model Rendering
Sung-eui Yoon, Enrico Gobbetti, David Kasik, and Dinesh Manocha
2008

High Dynamic Range Video
Karol Myszkowski, Rafal Mantiuk, and Grzegorz Krawczyk
2008

GPU-Based Techniques for Global Illumination Effects
László Szirmay-Kalos, László Szécsi, and Mateu Sbert
2008

High Dynamic Range Image Reconstruction
Asla M. Sá, Paulo Cezar Carvalho, and Luiz Velho
2008

High Fidelity Haptic Rendering
Miguel A. Otaduy and Ming C. Lin
2006

A Blossoming Development of Splines
Stephen Mann
2006

Mathematical Basics of Motion and Deformation in Computer Graphics: Second Edition

Ken Anjyo and Hiroyuki Ochiai

ISBN: 978-3-031-01464-2   paperback
ISBN: 978-3-031-02592-1   ebook

DOI 10.1007/978-3-031-02592-1

A Publication in the Springer series
SYNTHESIS LECTURES ON VISUAL COMPUTING: COMPUTER GRAPHICS, ANIMATION, COMPUTATIONAL PHOTOGRAPHY, AND IMAGING

Lecture #27
Series Editor: Brian A. Barsky, *University of California, Berkeley*
Series ISSN
Print 2469-4215   Electronic 2469-4223

Mathematical Basics of Motion and Deformation in Computer Graphics

Second Edition

Ken Anjyo
OLM Digital, Inc.

Hiroyuki Ochiai
Kyushu University

SYNTHESIS LECTURES ON VISUAL COMPUTING: COMPUTER GRAPHICS, ANIMATION, COMPUTATIONAL PHOTOGRAPHY, AND IMAGING #27

ABSTRACT

This synthesis lecture presents an intuitive introduction to the mathematics of motion and deformation in computer graphics. Starting with familiar concepts in graphics, such as Euler angles, quaternions, and affine transformations, we illustrate that a mathematical theory behind these concepts enables us to develop the techniques for efficient/effective creation of computer animation.

This book, therefore, serves as a good guidepost to mathematics (differential geometry and Lie theory) for students of geometric modeling and animation in computer graphics. Experienced developers and researchers will also benefit from this book, since it gives a comprehensive overview of mathematical approaches that are particularly useful in character modeling, deformation, and animation.

KEYWORDS

motion, deformation, quaternion, Lie group, Lie algebra

Contents

Preface

In the computer graphics community, many technical terms, such as Euler angle, quaternion, and affine transformation, are fundamental and quite familiar words, and have a pure mathematical background. While we usually do not have to care about the deep mathematics, the graphical meaning of such basic terminology is sometimes slightly different from the original mathematical entities. This might cause misunderstanding or misuse of the mathematical techniques. Or, if we have just a bit more curiosity about pure mathematics relevant to computer graphics, it should be easier for us to explore a new possibility of mathematics in developing a new graphics technique or tool.

This volume thus presents an intuitive introduction to several mathematical basics that are quite useful for various aspects of computer graphics, focusing on the fundamental procedures for deformation and animation of geometric objects, and curve/surface editing. The objective of this book, then, is to fill the gap between the original mathematical concepts and the practical meanings in computer graphics without assuming any prior knowledge of pure mathematics. We then restrict ourselves to the mathematics for matrices, while we know there are so many other mathematical approaches far beyond matrices in our graphics community. Though this book limits the topics to matrices, we hope you can easily understand and realize the power of mathematical approaches. In addition, this book demonstrates our ongoing work, which benefits from the mathematical formulation we develop in this book.

This book is an extension of our early work that was given as SIGGRAPH Asia 2013 and SIGGRAPH 2014 courses. The exposition developed in this book has greatly benefited from the advice, discussions, and feedback of a lot of people. The authors are very much grateful to Shizuo Kaji at Yamaguchi University and J.P. Lewis at Victoria University of Wellington, who read a draft of this book and gave many invaluable ideas. The discussions and feedback from the audience at the SIGGRAPH courses are also very much appreciated. Many thanks also go to Gengdai Liu and Alexandre Derouet-Jourdan at OLM Digital for their help in making several animation examples included in this book.

This work was partially supported by Core Research for Evolutional Science and Technology (CREST) program "Mathematics for Computer Graphics" of Japan Science and Technology Agency (JST). Many thanks especially to Yasumasa Nishiura at Tohoku University and Masato Wakayama at Institute of Mathematics for Industry, who gave long-term support to the authors.

The authors wish to thank Ayumi Kimura for the constructive comments and suggestions made during the writing of this volume. Thanks also go to Yume Kurihara for the cute illustrations. Last, but not least, the authors are immensely grateful to Brian Barsky, the editor of the Synthesis

Lectures on Compute Graphics and Animarion series, and Mike Morgan at Morgan & Claypool Publishers for giving the authors such an invaluable chance to publish this book in the series.

Ken Anjyo and Hiroyuki Ochiai
October 2014

Preface to the Second Edition

In this edition, we added an appendix where we derive several formulas for 3D rotation and deformation. We also incorporated a number of references, particularly relating to our SIGGRAPH 2016 course and its accompanying video. These additions will help readers to better understand the basic ideas developed in this book. We also resolved the mathematical notation inconsistency from the first edition, which makes this book more easily accessible.

We would like to thank the graduate students at Kyushu University who carefully went through the book with the second author in his seminar. Finally, we are very grateful to Ayumi Kimura who worked with us for the SIGGRAPH 2016 course and helped significantly with the editing of the second edition.

Ken Anjyo and Hiroyuki Ochiai
April 2017

Symbols and Notations

T_b	translation, 5
R_θ	2D rotation matrix, 6
$SO(2)$	2D rotation group (special orthogonal group), 7
$M(n, \mathbb{R})$	the set of square matrices of size n with real entries, 7
I, I_n	the identity matrix of size n, 7, 11, 30,
A^T	transpose of a matrix A, 7
$SE(2)$	2D motion group (the set of non-flip rigid transformations), 8
$O(2)$	2D orthogonal group, 9
$SO(n)$	special orthogonal group, 10
$O(n)$	orthogonal group, 10
$SO(3)$	3D rotation group, 11
$R_x(\theta)$	3D axis rotations, 11
$\mathbb{H}$	the set of quaternions, 3, 13
$\bar{q}$	conjugate of a quaternion, 13
$\mathrm{Re}(q)$	real part of a quaternion, 14
$\mathrm{Im}(q)$	imaginary part of a quaternion, 14
$\mathrm{Im}\,\mathbb{H}$	the set of imaginary quaternions, 14
$|q|$	the absolute value of a quaternion, 14
$\mathbb{S}^3$	the set of unit quaternions, 15
$\exp$	exponential map, 16, 29
$\mathrm{slerp}(q_0, q_1, t)$	spherical linear interpolation, 16
ε	dual number, 16
$M(2, \mathbb{H})$	the set of square matrices of size 2 with entries in $\mathbb{H}$, 16
$\check{\mathbb{C}}$	the set of anti-commutative dual complex numbers (DCN), 18
$E(n)$	rigid transformation group, 19
$SE(n)$	n-dimensional motion group, 20
$GL(n)$	general linear group, 23
$\mathrm{Aff}(n)$	affine transformations group, 23
$GL^+(n)$	general linear group with positive determinants, 23
$\mathrm{Aff}^+(n)$	the set of orientation-preserving affine transformations, 23
$\ltimes$	semi-direct product, 25
$\mathrm{Sym}^+(n)$	the set of positive definite symmetric matrices, 26
$\mathrm{Diag}^+(n)$	the set of diagonal matrices with positive diagonal entries, 27

SVD	Singular Value Decomposition, 28
$\exp(A)$	exponential of a square matrix, 29
$\mathbb{C}^{\times}$	the set of non-zero complex numbers, 31
$\mathfrak{gl}(n)$	Lie algebra of $GL(n)$, 33
$\mathfrak{so}(n)$	Lie algebra of $SO(n)$, 33
$\mathfrak{sl}(n)$	Lie algebra of $SL(n)$, 33
$\mathfrak{aff}(n)$	Lie algebra of $\text{Aff}(n)$, 34
$\mathfrak{se}(n)$	Lie algebra of $SE(n)$, 34
$[A, B]$	Lie bracket, 34
J_x, J_y, J_z	basis of $\mathfrak{so}(3)$, 37
$\log$	logarithmic map (logarithm), 31, 42
A^L, A^P, A^E	interpolant, 42
E^P, E^F, E^S, E^R	error functions, 49
$\|\cdot\|_F$	Frobenius norm of a matrix, 49
$\mathfrak{se}(3)$	Lie algebra of $SE(3)$, 53
$\mathfrak{sym}(3)$	the set of symmetric matrices of size three, 53
ϕ, ψ	map between $\text{Aff}^+(3)$ and a vector space, 53
ι	embedding $M(3, \mathbb{R}) \to M(4, \mathbb{R})$, 54
$\hat{R}, \hat{X}$	element of $SE(3)$ and $\mathfrak{se}(3)$, 54
∇	gradient, 58
Δ	Laplacian, 58
div	divergence, 58
$\partial\Omega$	boundary, 58

CHAPTER 1

Introduction

ORGANIZATION

In the latter half of this chapter we give a very rough sketch of several mathematical concepts that will reappear throughout this book.

In Chapters 2 and 3, we describe rigid and non-rigid transformations, while explaining the basic definitions regarding the *matrix group*. We thereafter show that Lie theoretic framework gives us comprehensive understanding of affine transformations, quaternions, and dual quaternions in Chapters 4 and 5. The Lie theoretic approach is also successfully applied to parametrization issues in Chapters 6 and 7, where we provide several useful recipes for rigid motion description and global deformation, along with our recent work. Finally in Chapter 8, we show a list of further readings, suggesting the power of mathematical approaches in graphics far beyond the present volume.

Here are a few additional notes that make this book easy to read and more enjoyable. First there are several colored columns in this book, which give brief, interesting stories of mathematicians or deeper explanations of the mathematical concepts in the body text. You may skip them at the first reading, but they will give you good guidance for your further study. Second, in this book, a point in Euclidean space is given as a row vector, whereas many geometric transformations are described with matrices. The action of a matrix to a vector then means multiplication from the left. As you may know, OpenGL takes the same manner of matrix multiplication, whereas DirectX does not.

A FEW MATHEMATICAL CONCEPTS

In this section, we therefore take a brief look at the original mathematical concepts related with matrices. These will be useful when we reuse or extend the basic ideas behind those concepts that are usually not well described in the computer graphics literature.

However, except the concept of *group*, we won't mention their rigorous definitions in mathematics. Rather we would like to describe the crude introduction of the mathematical concepts that are important even in computer graphics. A bit more precise definitions of them may also be given in later chapters. It would, however, be more important to think of why those mathematical concepts are useful in our graphics context, rather than learning deeply their rigorous mathematical entities.

GROUP

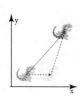

Let G be a set associated with an operation ".". If the pair $(G, \cdot)$ satisfies the following properties, then it is called a *group*. Or we would call G itself a group:

1. For any $a, b \in G$, the result of the operation, denoted by $a \cdot b$, also belongs to G.

2. For any a, b and $c \in G$, we have $a \cdot (b \cdot c) = (a \cdot b) \cdot c$.

3. There exists an element $e \in G$, such that $e \cdot a = a \cdot e = a$, for any element $a \in G$. (The element is then called the *identity* of G).

4. For each $a \in G$, there exists an element $b \in G$ such that $a \cdot b = b \cdot a = e$, where e is the identity. (The element b is then called the inverse of a.)

As usual, $\mathbb{R}$ and $\mathbb{C}$ denote the set of all real numbers and the set of all complex numbers, respectively. $\mathbb{R}$ or $\mathbb{C}$ is then a group with addition (i.e., the operation "." simply means $+$), and called *commutative*, since $a + b = b + a$ holds for any element a, b of $\mathbb{R}$ or $\mathbb{C}$. In the following sections, we'll see many groups of matrices. For example, the set of all invertible square matrices constitutes a group with composition as its group operation. The group consisting of the invertible matrices with size n is called the general linear group of order n, and will be denoted by $GL(n, \mathbb{R})$ or $GL(n, \mathbb{C})$.

LIE GROUP AND LIE ALGEBRA

A Lie group is defined to be a smooth manifold with a group structure. But we never mind what is a manifold (i.e., locally it is diffeomorphic to n-dimensional open disk). In applications, a matrix group, that is, a group consisting of matrices, like $GL(n, \mathbb{R})$ for instance, are enough to be considered as a Lie group. The totality of quaternions of unit length constitutes another Lie group. Although there is a general definition of Lie algebra, in this book we restrict ourselves to consider the Lie algebra associated with a Lie group. We then define the Lie algebra as a tangent space at the identity of the Lie group. In this sense, the Lie algebra can be considered as a linear approximation of the Lie group, which will be more explicitly described for the matrix groups in the following chapters.

QUATERNION

The original definition of quaternion by William Hamilton seems a bit different from the one we use in graphics. In 1835 he justified calculation for complex numbers $x + iy$ as those for ordered pairs of two real numbers (x, y). As is well known, complex numbers can express 2D rotations. This motivates many mathematicians to find a generalization of numbers which can describe 3D rotations. In 1843 he finally discovered it, referring to the totality of those numbers as *quaternions*. In this book, the set of quaternions is denoted by $\mathbb{H}$, and expressed as $\mathbb{H} = \mathbb{R} + \mathbb{R}i + \mathbb{R}j + \mathbb{R}k$, where we introduce the three numbers i, j and k satisfying the following rules:

$$i^2 = j^2 = k^2 = -1$$
$$ij = -ji = k.$$

$\mathbb{H}$ is then called an *algebra* or *field* (see [Ebbinghaus1991] for more details). We also note that, as shown in the above rules, it is not commutative. A few more alternative definitions of quaternions will also be given later for our graphics applications. In particular we'll see how 3D rotations can be represented with quaternions of unit length.

DUAL QUATERNION

In 1873, as a further generalization of quaternions, William K. Clifford obtained the concept called *biquaternions*, which is now known as a *Clifford algebra*. The concept of dual quaternions, which is another Clifford algebra, was also introduced in the late 19th century. A dual quaternion can be represented with $q = q_0 + q_\varepsilon \varepsilon$, where $q_0, q_\varepsilon \in \mathbb{H}$ and ε is the dual unit (i.e., ε commutes with every element of the algebra, while satisfying $\varepsilon^2 = 0$). We'll see later how rigid transformations in 3D space can be represented with dual quaternions of unit length.

CHAPTER 2

Rigid Transformation

In physics, a rigid body means as an object which preserves the distances between any two points of it with or without external forces over time. So describing rigid transformation (or rigid motion) means finding the *non-flip* congruence transformations parametrized over time. For a rigid body X, an animation (or a motion) $X(t)$ indexed by a time parameter t can be described by a series of rigid transformations $S(t)$ with $X(t) = S(t)X(0)$, instead of dealing with the positions of all the particles consisting of X. In the following sections, a non-flip congruence transformation may also be called a rigid transformation. The totality of the non-flip rigid transformations constitutes a *group*, which will be denoted by $SE(n)$, where n is the dimension of the world where rigid bodies live ($n = 2$ or 3). So let's start with 2D translation, a typical rigid transformation in $\mathbb{R}^2$.

2.1 2D TRANSLATION

A translation T_b by a vector $b \in \mathbb{R}^2$ gives a rigid transformation in 2D. The composition of two translations and the inverse of a translation, which is denoted by T_b^{-1}, are also translations:

$$T_b \cdot T_{b'} = T_{b+b'}, \quad T_b^{-1} = T_{-b}.$$

This can be rephrased as the totality of translations forms a *group* (recall Chapter 1). Moreover, they satisfy also

$$T_b \cdot T_{b'} = T_{b'} \cdot T_b.$$

This means that the totality of translations forms a *commutative* group. This property is illustrated in Figure 2.1. A commutative group is also called *abelian group* named after Niels Abel. The totality of 2D translations are denoted by $\mathbb{R}^2$, as is the two-dimensional vector space.

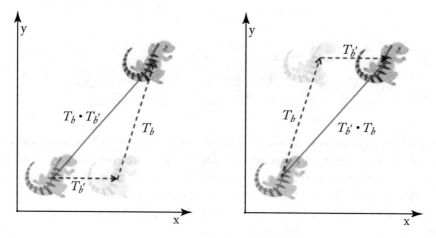

Figure 2.1: Example of groups—commutative group.

Niels Henrik Abel (1802–1829)
Norway mathematician. In his 28-year life, he gave a lot of important insight, which now has become a mathematical notion, such as Abelian groups, Abelian integral, Abelian functions, named after him. Also, the Abel prize was founded in 2002, the two-hundredth anniversary of Abel's birth. This prize is awarded to one or few outstanding mathematicians each year with six million kroner (approx. one million dollars).

2.2   2D ROTATION

A rotation in 2D centered at the origin (illustrated as in Figure 2.2) is then expressed by a matrix

$$R_\theta = \begin{pmatrix} \cos\theta & -\sin\theta \\ \sin\theta & \cos\theta \end{pmatrix}. \tag{2.1}$$

Note that the angle $\theta \in \mathbb{R}$ is not uniquely determined. To be more precise, two matrices R_θ and $R_{\theta'}$ give the same rotation if and only if $\theta - \theta'$ is an integer multiple of 2π. The compositions of

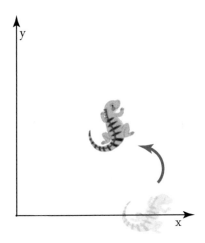

Figure 2.2: 2D rotation.

two rotations and the inverse of a rotation are again rotations:

$$R_\theta R_{\theta'} = R_{\theta+\theta'}, \quad R_\theta^{-1} = R_{-\theta}.$$

Here R_θ^{-1} denotes the inverse of R_θ. The totality of the rotations in 2D forms a *group* (also recall the definition of group in Chapter 1). It is denoted by

$$SO(2) = \{R_\theta \mid \theta \in \mathbb{R}\}. \tag{2.2}$$

We also write as

$$SO(2) = \{A \in M(2, \mathbb{R}) \mid AA^T = I, \det A = 1\}, \tag{2.3}$$

where $M(2, \mathbb{R})$ is the set of square matrices of size two, I is the identity matrix, and det is the determinant. The transpose[1] of a matrix A is denoted by A^T. A matrix $A \in M(2, \mathbb{R})$ is a rotation matrix if and only if the column vectors $u, v \in \mathbb{R}^2$ of A form an orthonormal basis and the orientation from u to v is counter-clockwise. This means that a rotation matrix sends any orthonormal basis with the positive orientation to some orthonormal basis with the positive orientation.

The result of the composition of several rotations in 2D is not affected by the order. This fact comes from the commutativity; $R_\theta R_{\theta'} = R_{\theta'} R_\theta$. Note that this is never true for 3D or a higher dimensional case.

---

[1]There are several manners to write a transpose of a matrix; A^T is rather popular but we will use [e.g., in Equation (5.2)] the notation A^t to express the t-th power of a matrix A for a real number t, so that we want to avoid this conflict. Another choice to write the transpose of a matrix A will be $^t\!A$.

2.3 2D RIGID TRANSFORMATION

The translation T_b does not preserve the origin (unless b is a zero vector), so it cannot be naturally expressed by a 2×2 matrix. A *homogeneous expression*[2] can express a translation T_b in 3×3 matrices:

$$\begin{pmatrix} 1 & 0 & b_1 \\ 0 & 1 & b_2 \\ 0 & 0 & 1 \end{pmatrix} \begin{pmatrix} x \\ y \\ 1 \end{pmatrix} = \begin{pmatrix} x' \\ y' \\ 1 \end{pmatrix}. \tag{2.4}$$

A successive operation (illustrated in Figure 2.3) of a rotation R_θ and a translation T_b maps a column vector $\mathbf{x} = (x, y)^T \in \mathbb{R}^2$ to $T_b(R_\theta(\mathbf{x})) = R_\theta \mathbf{x} + b$. In a homogeneous expression, it is written as

$$\begin{pmatrix} R_\theta & b \\ 0 & 1 \end{pmatrix} \begin{pmatrix} \mathbf{x} \\ 1 \end{pmatrix} = \begin{pmatrix} \mathbf{x}' \\ 1 \end{pmatrix}. \tag{2.5}$$

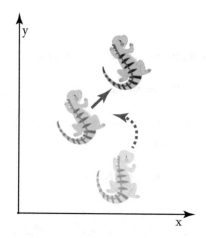

Figure 2.3: 2D rigid transformation.

Conversely, any orientation-preserving rigid transformation is uniquely written of this form (2.5). We denote by $SE(2)$ the set of non-flip rigid transformations. Then

$$SE(2) = \left\{ \begin{pmatrix} R & b \\ 0 & 1 \end{pmatrix} \in M(3, \mathbb{R}) \mid R \in SO(2), b \in \mathbb{R}^2 \right\}. \tag{2.6}$$

If we reverse the order of composition of a translation T_b and a rotation R_θ, the result $R_\theta T_b$ is different from $T_b R_\theta$. To be more precise, we have

$$R_\theta T_b = T_{b'} R_\theta \quad \text{with} \quad b' = R_\theta(b). \tag{2.7}$$

---

[2]See also Section 2.11 for homogeneous expression.

Note that the rotation component R_θ does not depend on the order of composition, while the translation part T_b or $T_{b'}$ does. This fact can be rephrased as the rigid transformation group in 2D is the *semi-direct product*[3] of the rotation group with the translation group, and can be denoted by $SE(2) = SO(2) \ltimes \mathbb{R}^2$ for short. Note that Equation (2.7) can be written as

$$R_\theta T_b R_{-\theta} = T_{b'}. \tag{2.8}$$

This property is rephrased as the group of translations is a *normal*[4] subgroup of the rigid transformation group, and it is denoted by $\mathbb{R}^2 \lhd SE(2)$. In Section 2.2, we discuss rotations centered at the origin. In general, the rotation with angle θ centered at $b \in \mathbb{R}^2$ can be expressed by

$$T_b R_\theta T_{-b}. \tag{2.9}$$

This bears a resemblance to (2.8), but the role of rotations and translations are reversed.

2.4   2D REFLECTION

A reflection (flip) with respect to a line $y = (\tan \theta)x$ through the origin can be expressed as

$$R_\theta \begin{pmatrix} 1 & 0 \\ 0 & -1 \end{pmatrix} R_{-\theta} = \begin{pmatrix} \cos 2\theta & \sin 2\theta \\ \sin 2\theta & -\cos 2\theta \end{pmatrix}. \tag{2.10}$$

A reflection is orientation-reversing transformation which preserves the shape. The determinant of the matrix (2.10) is -1. The composition of two reflections is a rotation, and the resulting rotation depends on the order of compositions of reflections:

$$\begin{pmatrix} \cos 2\theta & \sin 2\theta \\ \sin 2\theta & -\cos 2\theta \end{pmatrix} \begin{pmatrix} \cos 2\theta' & \sin 2\theta' \\ \sin 2\theta' & -\cos 2\theta' \end{pmatrix} = R_{2\theta - 2\theta'}. \tag{2.11}$$

The totality of rotations and reflections form an *orthogonal group* of size two, which is defined by

$$O(2) = \{A \in M(2, \mathbb{R}) \mid AA^T = I_2\}. \tag{2.12}$$

The totality of rotations has been denoted by

$$SO(2) = \{A \in O(2) \mid \det(A) = 1\}. \tag{2.13}$$

In the terminology in Section 3.2, $SO(2)$ is a *normal subgroup* of $O(2)$. Any reflection is not considered to be a motion, since it cannot be continuously connected with the identity transformation. In other words, $O(2)$ is not connected while $SO(2)$ is connected. Note that a *connected component* is like an island, illustrated in Figure 2.4. With this terminology, $SO(2)$ is a connected component. It is known that any two 2D reflections are continuously connected. This means

---

[3]This notion is explained in Section 3.2.
[4]This terminology will be explained in Section 3.2.

Figure 2.4: Connected components.

that the set of 2D reflections is a different connected component of $O(2)$ from $SO(2)$. The same holds for an arbitrary dimension n. That is, $SO(n)$ is connected, and $O(n)$ has two connected components. The set of the elements of $O(n)$ whose determinants are -1 is the other connected component of $SO(n)$. Moreover, for any two elements $g, h \in O(n)$ with $g, h \notin SO(n)$, we have $gh \in SO(n)$. This fact is rephrased as the *index* of the subgroup $SO(n)$ in $O(n)$ is two, and denoted by $[O(n) : SO(n)] = 2$.

2.5   3D ROTATION: AXIS-ANGLE

So far, we have discussed 2D rotations and flips. We now consider 3D rotations.

Given a unit vector $\mathbf{u} \in \mathbb{R}^3$ and an angle θ, the rotation with the axis $\mathbf{u}$ and the angle θ is, illustrated in Figure 2.5, given by

$$\mathbf{x} \mapsto R\mathbf{x} = \mathbf{x}' = (\cos\theta)\mathbf{x} + (\sin\theta)(\mathbf{u} \times \mathbf{x}) + (1 - \cos\theta)(\mathbf{u} \cdot \mathbf{x})\mathbf{u}. \tag{2.14}$$

This is called Rodrigues's rotation formula.

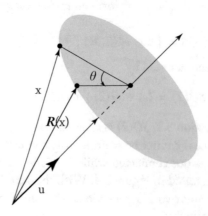

Figure 2.5: Axis angle representation.

Also in matrix notation

$$R = I + (\sin \theta) \begin{pmatrix} 0 & -u_3 & u_2 \\ u_3 & 0 & -u_1 \\ -u_2 & u_1 & 0 \end{pmatrix} + (1 - \cos \theta)(\mathbf{u}\mathbf{u}^T - I). \qquad (2.15)$$

See Section A.1 for more detail on Rodrigues formulas.

If we choose an orthonormal basis $\{\mathbf{u}, \mathbf{v}, \mathbf{w}\}$ of $\mathbb{R}^3$ with the right orientation, that is, $\mathbf{w} = \mathbf{u} \times \mathbf{v}$, then

$$\begin{aligned} \mathbf{u}' &= (\cos \theta)\mathbf{u} + (1 - \cos \theta)\mathbf{u} = \mathbf{u}, & (2.16) \\ \mathbf{v}' &= (\cos \theta)\mathbf{v} + (\sin \theta)(\mathbf{u} \times \mathbf{v}) = (\cos \theta)\mathbf{v} + (\sin \theta)\mathbf{w}, & (2.17) \\ \mathbf{w}' &= (\cos \theta)\mathbf{w} - (\sin \theta)\mathbf{v}. & (2.18) \end{aligned}$$

The rotation (2.14) is expressed as

$$R = \begin{pmatrix} u_1 & v_1 & w_1 \\ u_2 & v_2 & w_2 \\ u_3 & v_3 & w_3 \end{pmatrix} \begin{pmatrix} 1 & 0 & 0 \\ 0 & \cos \theta & -\sin \theta \\ 0 & \sin \theta & \cos \theta \end{pmatrix} \begin{pmatrix} u_1 & u_2 & u_3 \\ v_1 & v_2 & v_3 \\ w_1 & w_2 & w_3 \end{pmatrix}. \qquad (2.19)$$

Every 3D rotation turns out to be an element of special orthogonal group defined by

$$SO(3) = \{R \in M(3, \mathbb{R}) \mid RR^T = I_3, \det(R) = 1\}, \qquad (2.20)$$

where I_n denotes the identity matrix of size n. The converse is also true. In other words, every special orthogonal transformation in 3D is a rotation. (This fact is true only in 2D and 3D, and is never true for 4D or higher dimensions. That is, for $n > 3$, most of elements in $SO(n)$ have no rotation axis.) By this expression (2.20), the composition of two rotations is also a rotation. Along with the fact that the inverse of a rotation is also a rotation, we can say that the set of rotations is a *group*.

Note that two 3D rotations are not necessarily commutative, i.e., their compositions depend on the order of operations. For example, the successive operation of the rotation with respect to x-axis with angle $\pi/6$ and that with respect to z-axis with angle $\pi/4$ is different from their reversed ordered compositions (see Figure 2.6).

2.6   3D ROTATION: EULER ANGLE

The second method to express 3D rotations is so-called *Euler angle*. Any 3D rotation is a composition of three successive rotations along the coordinate axes:

$$\begin{aligned} &R_z(\theta_3) R_y(\theta_2) R_x(\theta_1) \\ &= \begin{pmatrix} \cos \theta_3 & -\sin \theta_3 & 0 \\ \sin \theta_3 & \cos \theta_3 & 0 \\ 0 & 0 & 1 \end{pmatrix} \begin{pmatrix} \cos \theta_2 & 0 & \sin \theta_2 \\ 0 & 1 & 0 \\ -\sin \theta_2 & 0 & \cos \theta_2 \end{pmatrix} \begin{pmatrix} 1 & 0 & 0 \\ 0 & \cos \theta_1 & -\sin \theta_1 \\ 0 & \sin \theta_1 & \cos \theta_1 \end{pmatrix}. \end{aligned} \qquad (2.21)$$

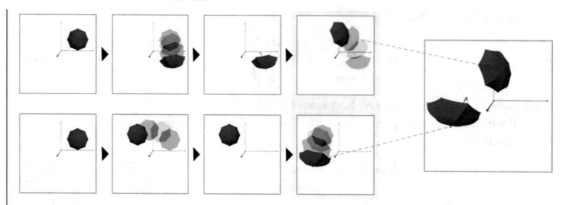

Figure 2.6: $R_x(\frac{\pi}{6})$ and $R_z(\frac{\pi}{4})$ are not commutative.

Each of the rotation matrices in the right-hand side can essentially be understood as a 2D rotation matrix, so that this method regards a 3D rotation as a composition of several (three) 2D rotations. This method respects the axis, so is not free from the choice of coordinates.

We have several versions of Euler angle representation. Here we employ successive *xyz*-rotations, but we may have six variations of three letters such as *yxz*, *zxy*, etc., as well as six variations of two letters such as *xyx*, *xzx*, etc. These variations look different, but are essentially the same. Within each class, we only need to exchange the name of coordinates. Between the classes, for example, the relations between *xyz*-Euler angle and *xyx*-Euler angle representations are given explicitly as follows. The identity

$$R_z(\theta)R_y(\frac{\pi}{2}) = R_y(\frac{\pi}{2})R_x(-\theta)$$

implies the formula

$$R_z(\theta_3)R_y(\theta_2)R_x(\theta_1) = R_y(\frac{\pi}{2})R_x(-\theta_3)R_y(\theta_2 - \frac{\pi}{2})R_x(\theta_1). \tag{2.22}$$

We understand this formula as follows: given a 3D rotation matrix with the *xyz*-Euler angle $(\theta_1, \theta_2, \theta_3)$, move by rotation $R_y(-\frac{\pi}{2})$, then we obtain another rotation matrix with the *xyx*-Euler angle $(\theta_1, \theta_2 - \frac{\pi}{2}, -\theta_3)$.

Every 3D rotation can be expressed by an Euler angle representation. Consider the *xyx*-Euler angle representation of a *z*-rotation

$$R_x(\theta_3)R_y(\theta_2)R_x(\theta_1) = R_z(\theta). \tag{2.23}$$

Then, for $0 < |\theta| < \pi$, we see that the solutions of this Equation (2.23) are $(\theta_1, \theta_2, \theta_3) = \pm(-\pi/2, \theta, \pi/2)$. This means that even if θ is close to zero, the angles θ_1 and θ_3 are far from zero, so that if we move θ continuously through zero, then its Euler angle $(\theta_1, \theta_2, \theta_3)$ may change discontinuously. Another intuitive explanation is that if we put $\theta_2 = 0$ in (2.23), then the left-hand side becomes $R_x(\theta_3 + \theta_1)$, so that only one-dimensional freedom remains though we expect two-parameter family. These phenomena are known as *Gimbal lock*, which is a demerit of this method. See [Ebbinghaus1991].

2.7   3D ROTATION: QUATERNION

Several Equivalent Definitions of Quaternions
Some deficiency of Euler angle is relaxed by using quaternions. We now here briefly recall the quaternions. For more detail, see the references [Shoemake1985, Watt1992, Hanson2006, Vince2011]. There are several ways of expressing quaternions:

(i) $\mathbb{H} = \mathbb{R} + \mathbb{R}i + \mathbb{R}j + \mathbb{R}k$, the real 4-dimensional vector space with the muliplication law $i^2 = j^2 = k^2 = -1, ij = k = -ji, jk = i = -kj, ki = j = -ki$.

(ii) The set of the real matrices of the form

$$\begin{pmatrix} a & -b & -c & -d \\ b & a & -d & c \\ c & d & a & -b \\ d & -c & b & a \end{pmatrix}. \tag{2.24}$$

(iii) The set of pairs (s, q) of real numbers s and real three-dimensional vectors $q \in \mathbb{R}^3$.

The relation between (i), (ii), and (iii) are given by $a + bi + cj + dk = sq$. Each of these realizations has some advantage and disadvantage. For example, in the picture (iii) the multiplication rule is inherited from the matrix multiplication, so that the associative law $q(q'q'') = (qq')q''$ is obvious, while in the picture (i) the multiplication rule is *defined* as $ij = k$, so that the associative law is non-trivial and to be examined (though it is easy and straightforward). The realization (iii) is most directly related with the description of 3D motion (e.g., 3D rotation by unit quaternion). In Section 2.9, we give two other equivalent definitions of quaternions. It should be emphasized that these five realizations are equivalent. We can choose and use an appropriate way according to each purpose from these equivalent realizations in order to understand, prove some formulae, and/or improve, make a code, etc. The notion of quaternions is a generalization of complex numbers. Most significant difference is the non-commutativity $qq' \neq q'q$, in general. However, quaternions share many nice properties, for example, quaternions form a ring (i.e., addition, multiplication), a vector space (i.e., multiplication by a real number), a field (i.e., every non-zero element has its inverse).

Unit Quaternions
The conjugate of a quaternion is defined by $\overline{a + bi + cj + dk} = a - bi - cj - dk \in \mathbb{R} + \mathbb{R}i + \mathbb{R}j + \mathbb{R}k = \mathbb{H}$. The real part of $q = a + bi + cj + dk$ is defined by a and denoted by $\text{Re}(q)$.

The imaginary part of $q = a + bi + cj + dk$ is defined by $bi + cj + dk$ and denoted by $\mathrm{Im}(q)$. We have a formula[5]

$$\mathrm{Re}(q) = \frac{1}{2}(q + \bar{q}), \quad \mathrm{Im}(q) = \frac{1}{2}(q - \bar{q}).$$

A quaternion is called *imaginary* if its real part is zero. The totality of imaginary quaternions is denoted by

$$\mathrm{Im}\,\mathbb{H} = \{bi + cj + dk \mid b, c, d \in \mathbb{R}\}.$$

The absolute value is denoted by $|q| = \sqrt{q\bar{q}} = \sqrt{a^2 + b^2 + c^2 + d^2}$ for $q = a + bi + cj + dk$. The set of unit quaternions

$$\mathbb{S}^3 = \{q \in \mathbb{H} \mid |q| = 1\} \tag{2.25}$$

is a group by a multiplication. (Note that there are several notations on the set of unit quaternions. Another notation is $Sp(1)$, the compact symplectic group of rank one.) Any unit quaternion is of the form

$$q = \cos\frac{\theta}{2} + (\sin\frac{\theta}{2})\mathbf{u}, \tag{2.26}$$

where $\mathbf{u}$ is a unit imaginary quaternion and a $\theta \in \mathbb{R}$. The multiplication $p \mapsto qpq^{-1} = qp\bar{q}$ gives an action of a unit quaternion q on $\mathbb{H}$.

What is an action?
For a group G and a set X, a map $G \times X \to X$ is called an *action* if it satisfies a variation of the associative law $(gh)x = g(hx)$ for $g, h \in G$ and $x \in X$ with an auxiliary condition $1x = x$ for all $x \in X$, where 1 is the identity element of the group G. We can simply say that G *acts* on X when the map $G \times X \to X$ is obvious from the context. For example, when $G = SO(2)$ and $X = \mathbb{R}^2$, the usual multiplication of a matrix and a vector gives an action of G on X.

This action preserves the imaginary quaternions

$$\mathrm{Im}\,\mathbb{H} = \{bi + cj + dk \mid b, c, d \in \mathbb{R}\}. \tag{2.27}$$

To be more explicit,

$$qpq^{-1} = (\cos\theta)\mathbf{p} + (\sin\theta)\mathbf{u} \times \mathbf{p} + (1 - \cos\theta)(\mathbf{u} \cdot \mathbf{p})\mathbf{u}, \tag{2.28}$$

which is equal to Rodrigues's Formula (2.14). We obtain a surjective group homomorphism. In general, for two groups G and H, a map $\phi : G \to H$ is called a group *homomorphism* if it satisfies $\phi(gh) = \phi(g)\phi(h)$ for all $g, h \in G$. This surjectivity means that every 3D rotation can be

---

[5]This is similar to the case of complex numbers, but not the same. In fact, for a complex number, the imaginary part is defined to be $\mathrm{Im}(z) = \frac{1}{2i}(z - \bar{z})$.

expressed by a unit quaternion, and the product of two unit quaternions expresses the composition of two rotations corresponding to these unit quaternions:

$$\mathbb{S}^3 \to SO(3). \tag{2.29}$$

This group homomorphism is illustrated in Figure 2.7. Imagine a point in $SO(3)$ and the corresponding point in $\mathbb{H}$ is moving. When a point in $SO(3)$ travels once, then the corresponding point in $\mathbb{H}$ does twice.

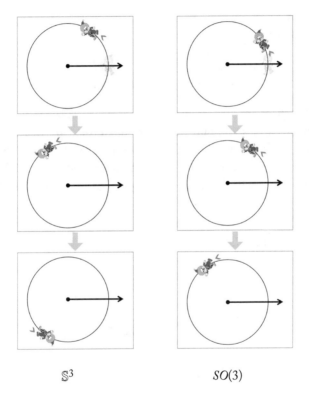

$\mathbb{S}^3$ $SO(3)$

Figure 2.7: Unit quaternion vs. rotation.

The transformation $q \mapsto qpq^{-1}$ is the identity transformation if and only if $q = \pm 1$. This situation can be expressed as a group *isomorphism*

$$\mathbb{S}^3/\{\pm 1\} \overset{\sim}{\to} SO(3). \tag{2.30}$$

This simply means that the rotation group $SO(3)$ can be identified with the group of unit quaternions with ± 1.

We give a meaning of (2.26) and its application to an interpolation by the exponential map. (A comprehensive treatment of exponential maps will be discussed later in Chapter 4.) The exponential map gives the surjective map

$$\exp : \operatorname{Im} \mathbb{H} \ni \theta\mathbf{u} \mapsto \cos\theta + (\sin\theta)\mathbf{u} \in \mathbb{S}^3. \tag{2.31}$$

For given $q_0, q_1 \in \mathbb{S}^3$, the *spherical linear interpolation* is given by

$$\operatorname{slerp}(q_0, q_1, t) = \frac{\sin((1-t)\theta)}{\sin\theta}q_0 + \frac{\sin(t\theta)}{\sin\theta}q_1, \tag{2.32}$$

where θ is given by the inner product $q_0 \cdot q_1 = \cos\theta$. This expression is explicit and fast, but does not explain why. Slerp satisfies

$$\operatorname{slerp}(q_0, q_1, t) = \operatorname{slerp}(1, q_1 q_0^{-1}, t)q_0, \tag{2.33}$$
$$\operatorname{slerp}(1, \exp(\theta\mathbf{u}), t) = \exp(t\theta\mathbf{u}), \tag{2.34}$$

reminding us that $q_0 \cdot q_1 = \cos\theta$ is equivalent to $\operatorname{Re}(q_1 q_0^{-1}) = \cos\theta$, and is equivalent to $q_1 q_0^{-1} = \cos\theta + (\sin\theta)\mathbf{u}$ with some unit imaginary quaternion $\mathbf{u}$. These two properties give a characterization of "slerp." The first equality is understood as an invariance under the right translation. The second equality is understood so that the interpolation $t \mapsto t\theta\mathbf{u}$ is chosen to be a linear interpolation of zero and $\theta\mathbf{u}$ in $\operatorname{Im} \mathbb{H}$. The Equations (2.31) and (2.34) can be illustrated in Figure 2.8.

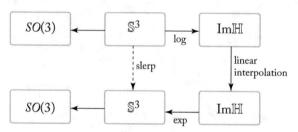

Figure 2.8: Slerp and log-exponential map.

2.8   DUAL QUATERNION

For a set R of numbers, its dual is defined to be $R + R\varepsilon$ with the rules $\varepsilon^2 = 0$ and $\varepsilon a = a\varepsilon$ for $a \in R$. This idea can be applied for $R = \mathbb{H}$ successfully describing the 3D rigid transformations, that is, rotations and translations. As are quaternions, several equivalent realizations of dual quaternions is useful.

(i) $\mathbb{H} + \mathbb{H}\varepsilon$ with $\varepsilon^2 = 0$.

(ii) A subalgebra of $M(2, \mathbb{H})$, consisting of matrices of the form $\begin{pmatrix} z & w \\ 0 & z \end{pmatrix}$.

We define a dual quaternion $z + w\varepsilon$ to be a unit dual quaternion if $|z| = 1$ and $\langle z, w \rangle = 0$, where $\langle \cdot, \cdot \rangle$ denotes the usual inner product on $\mathbb{R}^4 = \mathbb{H}$. For example, every unit quaternion is a unit dual quaternion, and an element of the form $1 + w\varepsilon$ with $w \in \operatorname{Im} \mathbb{H}$ is a unit dual quaternion. Conversely, every unit dual quaternion is written uniquely as a product of a unit quaternion z and a unit dual quaternion of the form $1 + w\varepsilon$ with $w \in \operatorname{Im} \mathbb{H}$. We identify a vector $(x, y, z)^T \in \mathbb{R}^3$ with the dual quaternion $1 + (xi + yj + zk)\varepsilon \in \mathbb{H} + \mathbb{H}\varepsilon$. Then, we define an action of a unit quaternion $q = z + w\varepsilon$ on $p = 1 + (xi + yj + zk)$ by $qp\bar{q}^*$, where $\bar{q}^* = \bar{z} - \bar{w}\varepsilon$. The action of unit quaternion z expresses a rotation in 3D, and the action of unit dual quaternion $q = 1 + w\varepsilon$ with $w \in \operatorname{Im} \mathbb{H}$ expresses a translation in 3D. See [Kavan2008] for detail. The semi-direct structure and the behavior of the exponential map on dual quaternions are understood both in the dual quaternion picture and matrix realization picture, as is the case of quaternions.

2.9   USING COMPLEX NUMBERS

Now we give a brief comment on complex numbers. Using the identification of $\mathbb{C}$ with $\mathbb{R}^2$ by $z = x + yi \leftrightarrow (x, y)^T$, a rigid transformation in 2D can be expressed as

$$\begin{pmatrix} \alpha & \beta \\ 0 & 1 \end{pmatrix} \begin{pmatrix} z \\ 1 \end{pmatrix} = \begin{pmatrix} z' \\ 1 \end{pmatrix}, \tag{2.35}$$

where $\alpha = e^{i\theta} = \cos\theta + i\sin\theta$, and $\beta = b_1 + b_2 i \in \mathbb{C}$. In other words, we have a realization of $SE(2)$ by a matrix with entries in complex numbers

$$SE(2) = \left\{ \begin{pmatrix} \alpha & \beta \\ 0 & 1 \end{pmatrix} \in M(2, \mathbb{C}) \mid \alpha, \beta \in \mathbb{C}, |\alpha| = 1 \right\}. \tag{2.36}$$

The reflection with respect to a line $y = (\tan\theta)x$ is represented in complex variables by

$$z \mapsto e^{i\theta}\overline{e^{-i\theta}z} = e^{2i\theta}\bar{z} = \overline{ze^{-2i\theta}}.$$

Note that the reflection line is also expressed as $\mathbb{R}e^{i\theta}$.

For quaternions, as well as the definitions (i)–(iii) listed in Section 2.7, we have other equivalent definitions using complex numbers.

(iv) $\mathbb{H} = \mathbb{C} + \mathbb{C}j$, the complex two-dimensional vector space with a multiplication rule. The expression (iv) is shorter than that of (i) in Section 2.7, while we note the fancy relation $wj = j\bar{w}$ for $w = \mathbb{C} = \mathbb{R} + \mathbb{R}i$.

(v) The set of complex matrices of the form

$$\begin{pmatrix} z & -w \\ \bar{w} & \bar{z} \end{pmatrix}. \tag{2.37}$$

The relation between (i)–(iii) in Section 2.7 and (iv), (v) is given by

$$a + bi + cj + dk = (a + bi) + (c + di)j = z + wj.$$

The multiplicativity $|qq'| = |q| \cdot |q'|$ of norms of quaternions follows from the property of determinant $\det(AB) = \det(A) \det(B)$ for the corresponding matrices A and B of the form (2.37).

2.10 DUAL COMPLEX NUMBERS

As is seen in Section 2.8, a unit dual quaternion can express an arbitrary 3D rigid transformation, say, every element in $SE(3)$. A 2D rigid transformation can be handled by regarding the plane embedded in $\mathbb{R}^3$ and using unit dual quaternions. We here introduce new numbers, called anti-commutative dual complex numbers, which will give more concise expression and faster computation of a 2D rigid transformation.

For two complex numbers $p_0, p_1 \in \mathbb{C}$, the combination $p_0 + p_1\varepsilon$ is called an anti-commutative dual complex number (DCN, for short) and denoted by $\check{\mathbb{C}}$. The multiplication of DCN is defined by

$$(p_0 + p_1\varepsilon)(q_0 + q_1\varepsilon) = (p_0q_0) + (p_1\bar{q}_0 + p_0q_1)\varepsilon. \tag{2.38}$$

From the definition, it follows that $\varepsilon^2 = 0$, and $(p_0 + p_1\varepsilon)(q_0 + q_1\varepsilon)$ may not be equal to $(q_0 + q_1\varepsilon)(p_0 + p_1\varepsilon)$. We also define the conjugate and absolute values of DCN by

$$\overline{p_0 + p_1\varepsilon} = \bar{p}_0 + p_1\varepsilon, \tag{2.39}$$
$$|p_0 + p_1\varepsilon| = |p_0|. \tag{2.40}$$

Then DCN satisfies usual associative and distributive laws,

$$(ab)c = a(bc), \quad a(b + c) = ab + ac, \quad (a + b)c = ac + bc, \tag{2.41}$$

so we can compute DCN as usual numbers. Similarly to the unit dual quaternion numbers, the unit anti-commutative complex numbers are of particular importance:

$$\check{\mathbb{C}}_1 = \{\hat{p} \in \check{\mathbb{C}} \mid |\hat{p}| = 1\} = \{e^{i\theta} + p_1\varepsilon \mid \theta \in \mathbb{R}, p_1 \in \mathbb{C}\}.$$

This is a group with the inverse

$$(e^{i\theta} + p_1\varepsilon)^{-1} = e^{-i\theta} - p_1\varepsilon.$$

A unit DCN $\hat{p}$ acts on $\mathbb{R}^2$ by identifying $v \in \mathbb{C} = \mathbb{R}^2$ with $1 + v\varepsilon \in \check{\mathbb{C}}$:

$$\hat{p}(1 + v\varepsilon)\overline{\hat{p}} = 1 + (p_0^2 v + 2p_0 p_1)\varepsilon,$$

that is, v is mapped to $p_0^2 v + 2p_0 p_1$. For example, when $p_1 = 0$, then $\hat{p} = p_0 = e^{i\theta}$ maps $v \in \mathbb{C}$ to $p_0^2 v = e^{2i\theta}v$, which is the rotation around the origin of degree 2θ. On the other hand, when $p_0 = 1$, the action corresponds to the translation by $2p_1 \in \mathbb{C} = \mathbb{R}^2$. Note that this gives a surjective group homomorphism $\varphi : \check{\mathbb{C}}_1 \to SE(2)$ whose kernel is $\{\pm 1\}$. In other words, any 2D rigid transformation corresponds to exactly two unit DCN's with opposite signs.

The following homomorphism

$$\check{\mathbb{C}} \ni p_0 + p_1 \varepsilon \mapsto p_0 + p_1 j \varepsilon \in \mathbb{H} + \mathbb{H}\varepsilon$$

is compatible with the involution and the conjugation, and preserves the norm. Furthermore if we identify $v = x + iy \in \mathbb{C}$ with $1 + (xj + yk)\varepsilon = 1 + vj\varepsilon$, the above map is commutative with the action. From this embedding, DCN is realized as a sub-ring of dual quaternion numbers. The relations among DQN, DCN and related groups are summarized in Figure 2.9. Here we add

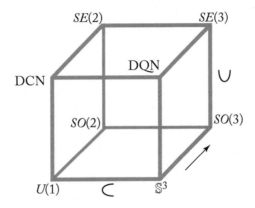

Figure 2.9: Dual quaternions and related groups.

some explanation. In each left-right edge, the object on the left is 2D, and that on the right is 3D. The left object is naturally contained in the right object. In each vertical edge, the top object contains the bottom object, and the top object is obtained from the bottom object by adding translations. In each front-behind edge, one element in the behind object corresponds exactly to two elements in the front object. We have a two-to-one surjective map from front to back.

The properties of DCN and its applications are given in [Matsuda2004].

2.11 HOMOGENEOUS EXPRESSION OF RIGID TRANSFORMATIONS

The homogeneous expression of nD rigid transformation group is

$$E(n) = \left\{ \begin{pmatrix} R & d \\ 0 & 1 \end{pmatrix} \in M(n+1, \mathbb{R}) \mid R \in O(n), d \in \mathbb{R}^n \right\}, \tag{2.42}$$

and its subgroup consisting of rigid motions (i.e., non-flip nD rigid transformations) is

$$SE(n) = \left\{ \begin{pmatrix} R & d \\ 0 & 1 \end{pmatrix} \in M(n+1, \mathbb{R}) \mid R \in SO(n), d \in \mathbb{R}^n \right\}. \tag{2.43}$$

As for a 2D case, we have it with (2.6).

Homogeneous coordinates

The term *homogeneous* comes from projective geometry. In a real vector space $\mathbb{R}^n$, two distinct lines meet at one point if they are not parallel. These two parallel lines are considered to meet at a point at infinity. If we add appropriate points of infinity to $\mathbb{R}^n$, we obtain a projective space, denoted by $\mathbb{P}^n(\mathbb{R})$. This space is explicitly realized as follows: we consider a non-zero vector with $(n+1)$-components

$$[z_1 : z_2 : \cdots : z_{n+1}].$$

If two such vectors are parallel, then we consider these two vectors to be same. In other words, if there is a non-zero number λ such that $z_1' = \lambda z_1, z_2' = \lambda z_2, \ldots, z_{n+1}' = \lambda z_{n+1}$, then we regard $[z_1 : z_2 : \cdots : z_{n+1}] = [z_1' : z_2' : \cdots : z_{n+1}']$. On the one hand, in the case $z_{n+1} \neq 0$, then we can take $\lambda = 1/z_{n+1}$ so that $[z_1 : z_2 : \cdots : z_{n+1}] = [z_1' : z_2' : \cdots : z_n' : 1]$. In this manner, an $(n+1)$-vector with $z_{n+1} \neq 0$ is regarded as a usual n-vector in $\mathbb{R}^n$. The coordinates $[z_1 : z_2 : \cdots : z_{n+1}]$ are called *homogeneous* coordinates since the values of coordinates have some homogenuity, while the classical coordinates $(x_1, \ldots, x_n) = [x_1 : \cdots : x_n : 1]$ are called *inhomogeneous* coordinates. On the other hand, in the case $z_{n+1} = 0$ an $(n+1)$-vector $[z_1 : z_2 : \cdots : z_{n+1}]$ gives an extra element other than $\mathbb{R}^n$, which is considered to be an element at infinity.

Now we consider the symmetry group of the projective spaces. A projective linear transformation on the projective space $\mathbb{P}^n(\mathbb{R})$ turns out to be represented by a multiplication of a regular matrix of size $(n+1)$:

$$\begin{pmatrix} a_{11} & \cdots & a_{1n} & a_{1n+1} \\ \vdots & & \vdots & \vdots \\ a_{n1} & \cdots & a_{nn} & a_{nn+1} \\ a_{n+11} & \cdots & a_{n+1n} & a_{n+1n+1} \end{pmatrix} \begin{pmatrix} z_1 \\ \vdots \\ z_n \\ z_{n+1} \end{pmatrix}. \tag{2.44}$$

A projective transformation preserves $\mathbb{R}^n$ if and only if $a_{n+11} = \cdots = a_{n+1n} = 0$. In such a case, since $a_{n+1n+1} \neq 0$ and a scalar multiple gives the same transformation, we may assume that $a_{n+1n+1} = 1$. A matrix of this form gives a homogeneous expression of an affine transformation of $\mathbb{R}^n$

$$\begin{pmatrix} a_{11} & \cdots & a_{1n} & a_{1n+1} \\ \vdots & & \vdots & \vdots \\ a_{n1} & \cdots & a_{nn} & a_{nn+1} \\ 0 & \cdots & 0 & 1 \end{pmatrix} \begin{pmatrix} x_1 \\ \vdots \\ x_n \\ 1 \end{pmatrix}. \tag{2.45}$$

CHAPTER 3

Affine Transformation

In this chapter, we discuss affine transformations, i.e., matrices for deformations.

3.1 SEVERAL CLASSES OF TRANSFORMATIONS

GL is not a graphic library (joke) but the general linear group, consisting of invertible linear transformations on $\mathbb{R}^n$. These are usually expressed in terms of square matrices with non-zero determinants:

$$GL(n) = GL(n, \mathbb{R}) = \{A \in M(n, \mathbb{R}) \mid \det(A) \neq 0\}. \tag{3.1}$$

An affine transformation is a map on $\mathbb{R}^n$, which maps every line to a line. These are usually expressed by a pair of an invertible square matrix and a vector in $\mathbb{R}^n$. The matrix shows the linear transformation and the vector does the translation. The set of all affine transformations is written as Aff(n). This group can also be expressed in the set of invertible square matrices of size $(n + 1)$ i.e., Aff(n) $\subset GL(n + 1)$:

$$\text{Aff}(n) := \left\{ \begin{pmatrix} A & d \\ 0 & 1 \end{pmatrix} \middle| A \in GL(n), d \in \mathbb{R}^n \right\}. \tag{3.2}$$

This realization is called a *homogeneous* expression. The composition of homogeneous expressions is nothing but a multiplication of two matrices.

If the determinant of a matrix in $GL(n)$ or Aff(n) is negative, then the corresponding transformation changes the orientation of objects. We denote the set of orientation-preserving transformations by

$$GL^+(n) = \{A \in GL(n) \mid \det(A) > 0\}, \tag{3.3}$$

$$\text{Aff}^+(n) = \left\{ \begin{pmatrix} A & d \\ 0 & 1 \end{pmatrix} \middle| A \in GL^+(n), d \in \mathbb{R}^n \right\}. \tag{3.4}$$

We here summarize the inclusion relations of these sets of transformations in Figure 3.1.

Here we add some explanations. In each left-right edge, the object on the left is an index-two subgroup of the object on the right. The object on the left is *connected* while the object on the right is *disconnected*. In each vertical edge, the object on the top is the semi-direct product of the object on the bottom with the translation group $\mathbb{R}^n$. In each front-behind edge, the object on the front is the subset of rigid transformations of the object on the behind.

The motion group (or Euclidean motion group), which is denoted by $SE(n)$ in (2.43), is the set of transformations on $\mathbb{R}^n$ which preserve the length, angle, and the orientation. Each element

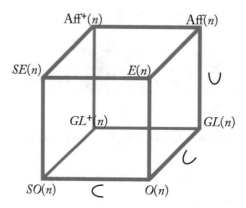

Figure 3.1: Inclusions of Lie groups.

of the motion group can be expressed as a pair of a rotation and a translation. The congruence group, denoted by $E(n)$, is the set of transformations which preserve the shape, but may change the orientation. A reflection is a typical example of congruence transformation. These classes of transformations are *groups*: The successive composition of transformations belongs to the same class of transformation, and the inverse transformation also does (see Chapter 1).

A bounded and closed subset of a vector space is called *compact* in terms of topology. The set of all rotations is bounded, while the set of all translations is unbounded. In these eight classes of groups, $SO(n)$ and $O(n)$ are compact, while other six classes of groups are not compact. The notion of *connected* and *arcwise connected* is equivalent in our cases. A maximal connected subset is called a connected component (see Section 2.4). We cannot continuously interpolate two elements in different connected components. For example, a flip and the identity cannot be interpolated in $GL(n)$. Actually $GL(n)$ consists of two connected components, each of which inlcudes the flip or the identity, respectively. On one hand, $SO(n)$, for instance, is connected, i.e, it has only one connected component which is itself.

Note that all these eight types of groups in Figure 3.1 are non-commutative for $n = 2, 3$ except for $SO(2)$.

Why should we consider so many groups? Do mathematicians like the complication? One may regard this large variety of Lie groups as a large variety of software in CG. Depending on the purpose, one may use a wide variety of computer languages and/or platforms, like C++, maya, python, MatLab, etc. This philosophy goes back to Klein's Erlangen program.

Felix Klein (1849–1925)
German mathematician. His research project was published in 1872 at Erlangen, and is called the Erlangen program. The slogan of the Erlangen program is that "symmetry classifies geometry." To be more precise, each class of geometry has the corresponding groups describing the symmetry, and one of the purposes of geometry is to describe its invariants. This idea is sufficient to understand Euclidean, affine and projective geometry, and their relations [Klein1926].

3.2 SEMIDIRECT PRODUCT

We see that the composition of a rotation, a translation, and the inverse rotation is another translation (2.8):

$$R_\theta T_b R_{-\theta} = T_{R_\theta(b)}. \tag{3.5}$$

In general, this fact is related to the notion of *normal subgroup* and *semi-direct* products of subgroups.

Let G be a group. The following concepts for G are used to describe the relations among the matrix group appeared in this book:

(i) A subset H of G is called a *subgroup* if H is closed under the composition and the inversion,

(ii) A subgroup H of G is called a *normal* subgroup if the composition $g \cdot h \cdot g^{-1}$ of any $h \in H$ and $g \in G$ belongs to H.

For example, the set $\mathbb{R}^3$ of translations is a normal subgroup of $SE(3)$, while the set $SO(3)$ of rotations centered at the identity is a subgroup of $SE(3)$ but not normal.

Let G be a group, H a subgroup of G, and K a normal subgroup of G. (For example, $G = SE(3), H = SO(3), K = \mathbb{R}^3$.) If the map

$$H \times K \ni (h, k) \mapsto hk \in G \tag{3.6}$$

is a bijective, then G is the *semi-direct product* of H and K, denoted by $H \ltimes K$. Note that $hkh^{-1} \in K$ for any $h \in H$ and $k \in K$, but is not necessarily equal to k. If both H and K are normal subgroups of G, and $hk = kh$ for all $h \in H$ and $k \in K$, and the map (3.6) is bijective, then G is called a *direct product* group $H \times K$. Note that $hk = kh$ if and only if $hkh^{-1} = k$. Motion groups

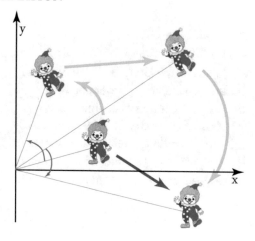

Figure 3.2: Translations are normal.

and affine transformation groups are typical examples of semi-direct product groups:

$$SE(n) = SO(n) \ltimes \mathbb{R}^n, \tag{3.7}$$
$$E(n) = O(n) \ltimes \mathbb{R}^n, \tag{3.8}$$
$$\text{Aff}^+(n) = GL^+(n) \ltimes \mathbb{R}^n, \tag{3.9}$$
$$\text{Aff}(n) = GL(n) \ltimes \mathbb{R}^n. \tag{3.10}$$

These decompositions can be interpreted as, for example, the translation part of a motion has its own meaning, which does not depend on the choice of coordinates and scaling, but a rotation part has some ambiguity, depending on the choice of the origin and that of the coordinates.

3.3   DECOMPOSITION OF THE SET OF MATRICES

Other than (semi-)direct product, several decompositions of matrices are widely used in computer graphics. Here we summarize the decompositions which will appear in the later sections.

3.3.1   POLAR DECOMPOSITION

Given a matrix $A \in GL^+(n)$, we have $A = RS$, where R is a rotation matrix and S is a positive definite symmetric matrix. The product map

$$SO(n) \times \text{Sym}^+(n) \ni (R, S) \mapsto RS \in GL^+(n) \tag{3.11}$$

is bijective.

Note that if we reverse the order

$$SO(n) \times \text{Sym}^+(n) \ni (R, S) \mapsto SR \in GL^+(n) \tag{3.12}$$

then it is still bijective, however it gives the different map. We also note that the set $\text{Sym}^+(n)$ is not a group; actually, the product of two elements in $\text{Sym}^+(n)$ is not necessarily symmetric.

3.3.2 DIAGONALIZATION OF POSITIVE DEFINITE SYMMETRIC MATRIX

Every positive definite symmetric matrix X is written as $X = RDR^T = RDR^{-1}$, where R belongs to $SO(n)$ and D is a diagonal matrix whose diagonal entries are all positive. Actually, the diagonal entries of D is the set of eigenvalues of given X. In general, the map

$$SO(n) \times \text{Diag}^+(n) \ni (R, D) \mapsto RDR^T \in \text{Sym}^+(n) \tag{3.13}$$

is surjective. Note that this map is not injective. If (R, D) and (R', D') expresses the same X, then there exists a permutation matrix P such that $D' = PDP^T$. Here a permutation matrix is, by definition, a matrix which has unique non-zero entry 1 in each row and each column. The inverse and the product of permutation matrices are also permutation matrices. Furthermore, if D has distinct diagonal entries, then P is unique. This means that the expression $X = RDR^T$ is not unique, but the freedom of choices exists only in the order of eigenvalues of X in the diagonal entries in D.

3.3.3 SINGULAR VALUE DECOMPOSITION (SVD)

Every matrix $A \in GL^+(2)$ can be written as $A = R_\alpha D R_\beta$, where $R_\alpha, R_\beta \in SO(2)$ and D is a diagonal matrix with positive diagonal entries (see Figure 3.3). In general, the product map

$$\begin{array}{ccc} SO(n) \times \text{Diag}^+(n) \times SO(n) & \to & GL^+(n) \\ (R', D, R) & \mapsto & R'DR \end{array} \tag{3.14}$$

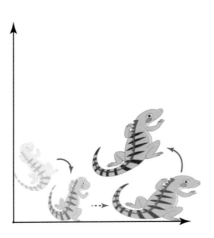

Figure 3.3: SVD; the curved arrows show rotations. The dotted line shows a directional dilation.

is surjective. This SVD is a combination of the polar decomposition and the diagonalization of positive symmetric matrix. In fact, if we denote by $A = R'S$ the polar decomposition of A, and by $S = RDR^T$, then the expression $A = (R'R)DR$ gives SVD. (Notice that $R'R \in SO(n)$.) This enables us to compute SVD by polar decomposition and the diagonalization of positive symmetric matrix. On the other hand, when we have SVD $A = R'DR$, then $A = (R'R)(R^T DR)$ gives polar decomposition of A.

Similarly, $O(n) \times \text{Diag}^+(n) \times O(n) \to GL(n)$ is also surjective. Note that SVD is called Cartan decomposition in mathematical literature (see, e.g., [Helgason1978]). In this point of view, the compactness of $SO(n)$ and commutativity of $\text{Diag}^+(n)$ is significant.

CHAPTER 4

Exponential and Logarithm of Matrices

Geometric transformations that we have described give a basic mathematical framework for geometric operations in computer graphics, such as rotation, shear, translation, and their compositions. Each affine transformation is then represented by a 4×4-homogeneous matrix (3.2) with usual operations: addition, scalar product, and product. While the product means the composition of the transformations, geometric meanings of addition and scalar product are not trivial. We often want to have a geometrically meaningful weighted sum (linear combination) of transformations, which is not an easy task. These kinds of practical demands therefore have inspired graphics researchers to explore new mathematical concepts and/or tools. Many works have been conducted in this direction, including skinning [Chaudhry2010, Lewis2000], cage-based deformation [Nieto2013], motion analysis and compression ([Alexa2002, Tournier2009], for instance).

In previous chapters we have described the geometric transformations through the mathematical concepts associated with groups, especially with the Lie group. We should then note that the mathematical viewpoint gives us a broader and more comprehensive scope of the various geometric transformations.

This chapter focuses a bit more on the Lie theoretic aspect of this scope. We introduce Lie algebra that associates the Lie group of matrices as a motion group. As will be demonstrated, the Lie algebra gives linear approximation of the Lie group, which allows us to use a powerful linear interpolation scheme in making dynamic motion and deformation.

4.1   DEFINITIONS AND BASIC PROPERTIES

We first consider a square matrix A, which is implicitly considered an element of a Lie group of matrices. The exponential of A is then defined as

$$\exp(A) = \sum_{k=0}^{\infty} \frac{1}{k!} A^k = I + A + \frac{1}{2}A^2 + \frac{1}{6}A^3 + \cdots, \tag{4.1}$$

where $A^0 = I$ is the identity matrix. We'll refer to (4.1) as the matrix exponential, for short. This is motivated by Taylor expansion of the usual exponential function

$$e^x = \sum_{k=0}^{\infty} \frac{1}{k!} x^k = 1 + x + \frac{1}{2}x^2 + \frac{1}{3!}x^3 + \cdots . \tag{4.2}$$

The series $\exp(A)$ converges for an arbitrary A rapidly, as does the usual exponential function. However, this infinite series expression is not so efficient for actual numerical computations. For a computation, we can use several useful properties: for diagonal matrices we have

$$\exp \begin{pmatrix} a & 0 & 0 \\ 0 & b & 0 \\ 0 & 0 & c \end{pmatrix} = \begin{pmatrix} e^a & 0 & 0 \\ 0 & e^b & 0 \\ 0 & 0 & e^c \end{pmatrix}, \tag{4.3}$$

and a rotation

$$\exp \begin{pmatrix} 0 & -\theta \\ \theta & 0 \end{pmatrix} = \begin{pmatrix} \cos \theta & -\sin \theta \\ \sin \theta & \cos \theta \end{pmatrix}. \tag{4.4}$$

We also see that the exponential image of a strictly upper-triangular matrix terminates into a finite sum. For example, for $l \in \mathbb{R}^n$, we have

$$\exp \begin{pmatrix} O & l \\ 0 & 0 \end{pmatrix} = \begin{pmatrix} I_n & l \\ 0 & 1 \end{pmatrix}. \tag{4.5}$$

Slightly more generally, for a strictly upper triangular matrix A of size n, we have $A^n = O$ and therefore the infinite series (4.1) terminates to a finite sum expression

$$\exp(A) = I + A + \frac{1}{2}A^2 + \cdots + \frac{1}{(n-1)!}A^{n-1}. \tag{4.6}$$

Significantly, we can understand Rodrigues's rotation formula (2.15) by using the matrix exponential. Every 3D rotation is expressed by

$$R = \exp(A) = \exp \begin{pmatrix} 0 & -u_3 & u_2 \\ u_3 & 0 & -u_1 \\ -u_2 & u_1 & 0 \end{pmatrix} = I_3 + \frac{\sin |\mathbf{u}|}{|\mathbf{u}|} A + \frac{1 - \cos |\mathbf{u}|}{|\mathbf{u}|^2} A^2, \tag{4.7}$$

where $|\mathbf{u}| = \sqrt{u_1^2 + u_2^2 + u_3^2}$ is the norm of a vector $\mathbf{u} = (u_1, u_2, u_3) \in \mathbb{R}^3$. We also see that $|\mathbf{u}|^2 = \frac{1}{2}\text{tr}(AA^T) = -\frac{1}{2}\text{tr}(A^2)$. The matrix R shows the rotation around the axis through $\mathbf{u}$, and with angle $|\mathbf{u}|$. In particular, if $|\mathbf{u}| \in 2\pi\mathbb{Z}$ then $R = I_3$, the identity matrix.

Coming back to the general situation, we always have the exponential law

$$\exp((s+t)A) = \exp(sA)\exp(tA) \quad \text{for all} \quad s, t \in \mathbb{R}, \quad A \in M(n, \mathbb{R}). \tag{4.8}$$

However a further generalization of the exponential law to a matrix case

$$\exp(A + B) \overset{?}{=} \exp(A)\exp(B) \quad \text{for} \quad A, B \in M(n, \mathbb{R}) \tag{4.9}$$

may <u>not</u> hold in general. For example, it is easy to see that three matrices $\exp(A + B)$, $\exp(A)\exp(B)$ and $\exp(B)\exp(A)$ are different, for $A = \begin{pmatrix} 0 & 0 \\ 1 & 0 \end{pmatrix}$ and $B = \begin{pmatrix} 0 & -1 \\ 0 & 0 \end{pmatrix}$. If we assume A and B are commutative, i.e., $AB = BA$, then we have an expected formula

$$\exp(A + B) = \exp(A)\exp(B) \quad \text{for} \quad A, B \in M(n, \mathbb{R}) \quad \text{with} \quad AB = BA. \tag{4.10}$$

In general, the matrix exponential has the conjugate invariance property

$$\exp(P^{-1}AP) = P^{-1}\exp(A)P. \tag{4.11}$$

The matrix exponential is by definition an infinite sum of matrices, but by this property (4.11) we can reduce the computation of the exponential map to the cases (4.3), (4.4), (4.6) so that we can avoid the infinite series for the computation of the exponential $\exp(A)$.

We now consider the inverse map of the exponential map: *logarithmic* map, or logarithm, for short. The logarithm might be defined to be the inverse of the exponential as is the case of the real scalar-valued function

$$\exp : \mathbb{R} \to \{y > 0\} = \mathbb{R}_{>0}. \tag{4.12}$$

However, we note that the logarithm of the complex exponential function

$$\exp : \mathbb{C} \to \{z \in \mathbb{C} \mid z \neq 0\} = \mathbb{C}^{\times}$$

is multi-valued. As is Euler's formula in complex numbers

$$\cos\theta + i\sin\theta = \exp(i\theta) = e^{i\theta}, \tag{4.13}$$

which gives an intimate connection between exponential functions and trigonometric functions, the exponential expression (4.4) of a rotation is not unique; $\theta + 2n\pi (n \in \mathbb{Z})$ gives the same rotation as θ. This feature makes the inverse of a matrix complicated; the logarithm of a matrix is thus multi-valued. As is the scalar valued function, the logarithm has a series expansion

$$\log(X) = \sum_{k=1}^{\infty} \frac{(-1)^{k-1}}{k}(X - I)^k$$

if the absolute values of all the eigenvalues of the matrix $X - I$ are less than 1. This function satisfies $\exp(\log(X)) = X$ as is expected. Similar to (4.11), the logarithm has a conjugate invariance $\log(P^{-1}XP) = P^{-1}\log(X)P$, and then the computation of $\log(X)$ is reduced to the case of scalar-valued functions and the case of triangular matrices, where infinite series terminates.

$$\log \begin{pmatrix} \cos\theta & -\sin\theta \\ \sin\theta & \cos\theta \end{pmatrix} = \begin{pmatrix} 0 & -\theta \\ \theta & 0 \end{pmatrix},$$

$$\log \begin{pmatrix} e^a & 0 & 0 \\ 0 & e^b & 0 \\ 0 & 0 & e^c \end{pmatrix} = \begin{pmatrix} a & 0 & 0 \\ 0 & b & 0 \\ 0 & 0 & c \end{pmatrix},$$

and if $(A - I)^m = O$ for some m, then

$$\log A = \sum_{k=1}^{m-1} \frac{(-1)^{k-1}}{k}(A - I)^k.$$

4.2   LIE ALGEBRA

The set $\mathfrak{so}(3)$ of skew-symmetric, that is, the transpose is its minus, 3×3 matrices is regarded as *Lie algebra* of $SO(3)$ (see Chapter 1 as well).

Figure 4.1: Lie algebra as a tangent space.

In this book, we consider a Lie group to consist of matrices, say, a matrix group. The Lie algebra of a Lie group is a linear approximation of a group at the identity. In general, the Lie algebra of a matrix group $G \subset GL(n)$ is defined to be the tangent space (see Figure 4.1) of G at the identity of G. Equivalently, the Lie algebra is the collection of elements in $M(n, \mathbb{R})$ of the form

$$\varphi'(0) = \frac{d}{dt}\varphi(t)\bigg|_{t=0} \tag{4.14}$$

for any curve $\varphi : \mathbb{R} \to G$ with $\varphi(0) = I$.

Sophus Lie (1842–1899)
Norwegian mathematician. He tried to control the continuous symmetry in geometry and differential equations by introducing its linearization. This idea is now regarded as a core of Lie Theory. Lie groups and Lie algebras are also named after him. Lie was a

close friend of F. Klein, and this communication led to mutual influence (see [Stubhaug2002]).

For example, let us compute the Lie algebra of the Lie group $SO(n)$. For any curve $\varphi :$ $\mathbb{R} \to SO(n)$, the image should satisfy $\varphi(t)\varphi(t)^T = I_n$. By differentiating with the Leibnitz rule, we obtain

$$0 = \frac{d}{dt}\varphi(t)\varphi(t)^T\bigg|_{t=0} = \frac{d\varphi(t)}{dt}\varphi(t)^T + \varphi(t)\frac{d\varphi(t)^T}{dt}\bigg|_{t=0} = A + A^T,$$

where we put $A = \varphi'(0)$ for short. This shows the Lie algebra of $SO(n)$ is the set of matrices A with $A^T = -A$. In general, the Lie algebra corresponding to a given Lie group is denoted by the corresponding "mathfrak" letters; for example, the Lie algebra of $SO(3)$ is denoted by $\mathfrak{so}(3)$, that of $GL(n)$ by $\mathfrak{gl}(n)$. We give several examples of Lie algebras.

(i) The Lie algebra of $GL(n)$ and $GL^+(n)$ is $\mathfrak{gl}(n) = M(n, \mathbb{R})$,

(ii) The Lie algebra of $O(n)$ and $SO(n)$ is $\mathfrak{so}(n) = \{A \in M(n, \mathbb{R}) \mid A^T = -A\}$.

(iii) The Lie algebra of the group of positive real numbers $\mathbb{R}_{>0}$ is $\mathbb{R}$. Note that the group law for $\mathbb{R}_{>0}$ is multiplicative, while that for $\mathbb{R}$ is additive.

(iv) The Lie algebra of $\mathbb{C}^\times$ is $\mathbb{C}$,

(v) The Lie algebra of $\mathbb{H}^\times = \{q \in \mathbb{H} \mid q \neq 0\}$ is $\mathbb{H}$,

(vi) The Lie algebra of the group $\mathbb{S}^3$ of unit quaternions is the set $\text{Im }\mathbb{H}$ of imaginary quaternions.

(vii) The special linear group is defined to be the set of volume-preserving linear maps; $SL(n, \mathbb{R}) = \{A \in GL(n, \mathbb{R}) \mid \det A = 1\}$. Its Lie algebra is $\mathfrak{sl}(n, \mathbb{R}) = \{A \in M(n, \mathbb{R}) \mid \text{tr}A = 0\}$, the set of traceless matrices.

(viii) The Lie algebra of the group of translations in 3D is $\mathbb{R}^3$.

The Lie algebra of a subgroup is a subspace of the Lie algebra. The Lie algebra of the direct product group $H \times K$ is $\mathfrak{h} \oplus \mathfrak{k}$, the direct sum of Lie algebras $\mathfrak{h}$ and $\mathfrak{k}$ of H and K. The Lie algebra of the semidirect product group $H \ltimes K$ is $\mathfrak{h} \oplus \mathfrak{k}$, the direct sum of Lie algebras $\mathfrak{h}$ and $\mathfrak{k}$ of H and K as a vector space. We give several examples of Lie algebras of direct or semi-direct product groups.

(ix) The Lie algebra of $\mathrm{Diag}^+(n) = \{X \mid$ diagonal matrices with positive diagonal entries$\}$ is $\mathfrak{diag}(n) = \{A \mid$ diagonal matrices$\}$.

(x) The Lie algebra of the affine transformation groups $\mathrm{Aff}^+(n)$ and $\mathrm{Aff}(n)$, defined in (3.2 and 3.4) is

$$\mathfrak{aff}(n) = \left\{ \begin{pmatrix} A & l \\ 0 & 0 \end{pmatrix} \mid A \in M(n, \mathbb{R}), l \in \mathbb{R}^n \right\}.$$

This fact will be used in Chapter 6 and 7.

(xi) The Lie algebra of $SE(n)$, defined in (2.43) and that of $E(n)$ is

$$\mathfrak{se}(n) = \left\{ \begin{pmatrix} A & l \\ 0 & 0 \end{pmatrix} \mid A \in \mathfrak{so}(n), l \in \mathbb{R}^n \right\}.$$

This will appear in Figure 5.3 for $n = 2$, and in Figure 7.2 for $n = 3$ of later sections.

(xii) The set of invertible dual quaternion numbers

$$\mathbb{H}^\times + \varepsilon\mathbb{H} = \{z + w\varepsilon \mid z, w \in \mathbb{H}, z \neq 0\} \tag{4.15}$$

is a Lie group by the multiplication. As a Lie group, it is a semidirect product group $\mathbb{H}^\times \ltimes \mathbb{H}$. The Lie algebra of $\mathbb{H}^\times + \varepsilon\mathbb{H}$ is $\mathbb{H} + \varepsilon\mathbb{H}$.

(xiii) The set of unit dual quaternion numbers

$$\{(z, w) \mid z \in \mathbb{S}^3, w \in \mathbb{H}, \langle z, w \rangle = 0\} \tag{4.16}$$

is a Lie group by the multiplication. Its Lie algebra is $\mathrm{Im}\,\mathbb{H} + \varepsilon\,\mathrm{Im}\,\mathbb{H}$.

For the Lie algebra $\mathfrak{gl}(n, \mathbb{R})$, we define a binary operation, called *Lie bracket*, by

$$[A, B] = AB - BA.$$

Abstract Lie algebra
We can easily check that this bracket operation satisfies the following properties:

(i) The bracket $[A, B]$ is bilinear, i.e., linear both in A and in B. To be explicit, $[\lambda A, B] = \lambda[A, B] = [A, \lambda B]$, $[A + C, B] = [A, B] + [C, B]$ and $[A, B + C] = [A, B] + [A, C]$,

(ii) The bracket is skew-symmetric, i.e., $[A, B] = -[B, A]$,

(iii) Jacobi identity: $[A, [B, C]] + [B, [C, A]] + [C, [A, B]] = 0$,

for all $A, B, C \in \mathfrak{g}$ and $\lambda \in \mathbb{R}$. Conversely(!), these three properties are exactly the definition of an (abstract) Lie algebra. In other words, a vector space equipped with a bracket operation with these properties is called a Lie algebra (see, e.g., [Serre1992]). In the case of sub vector spaces of matrix spaces, these three properties hold trivially, as is mentioned above. What should be confirmed to be a Lie algebra is the property: $[A, B] \in \mathfrak{g}$ for every $A, B \in \mathfrak{g}$.

4.3   EXPONENTIAL MAP FROM LIE ALGEBRA

So far, we have discussed the construction of Lie algebra from Lie groups. Now we discuss the converse direction: the construction of Lie groups from Lie algebra. Suppose $\mathfrak{g}$ is the Lie algebra of a group G. Then the exponential map gives a map from $\mathfrak{g}$ to G.

$$\exp : \mathfrak{g} \to G. \tag{4.17}$$

In fact, the exponential map gives a local diffeomorphism (= one-to-one, onto, smooth, and its inverse is also smooth) between a neighborhood of the origin of the vector space and a neighborhood of the identity of the group (the set of transformations). In general, the exponential map is not necessarily injective, nor surjective. We give several examples of exponential maps. Here the number is taken from the Lie algebra examples in the Section 4.2.

(iii) $\exp : \mathbb{R} \to \mathbb{R}_{>0}$ is a bijective(= one-to-one onto) map (4.12).

(ii) $\exp : \mathfrak{so}(2) \to SO(2)$ is surjective(= onto), but not injective(= one-to-one). The explicit form is given in (4.4).

(vi) $\exp : \operatorname{Im} \mathbb{H} \to \mathbb{S}^3$ is surjective, but not injective. The explicit form is given in (2.31).

(v) $\exp : \mathbb{H} \to \mathbb{H}^{\times}$ is surjective, but not injective. The explicit form is given by using the case (vi) of unit quaternions. For $a \in \mathbb{R}$ and $z \in \operatorname{Im} \mathbb{H}$, we have $\exp(a + z) = e^{a} \exp(z)$, where $\exp(z)$ is given in (vi).

(xii) The explicit form of the exponential map $\exp : \mathbb{H} + \varepsilon\mathbb{H} \to \mathbb{H}^{\times} + \varepsilon\mathbb{H}$ is given in Appendix of [Kavan2008].

(xi) The explicit form of the exponential map $\exp : \mathfrak{se}(3) \to SE(3)$ for 3D motion group is given in (7.3), (7.4) and (7.5).

(ix) The exponential map (4.3) gives a bijection

$$\exp : \mathfrak{diag}(n) \to \operatorname{Diag}^{+}(n). \tag{4.18}$$

The exponential map also induces a bijection

$$\exp : \mathfrak{sym}(n) \to \operatorname{Sym}^{+}(n), \tag{4.19}$$

where

$$\begin{aligned}\operatorname{Sym}^{+}(n) &= \{X \in M(n, \mathbb{R}) \mid \text{symmetric, positive definite}\}, \\ \mathfrak{sym}(n) &= \{A \in M(n, \mathbb{R}) \mid \text{symmetric}\}.\end{aligned}$$

Note that $\operatorname{Sym}^{+}(n)$ is *not* a group, and that $\mathfrak{sym}(n)$ is *not* a Lie algebra. The relation between (4.19) and (4.18) has been suggested in (3.13): $\exp(RXR^{T}) = R\exp(X)R^{T}$ for $R \in SO(n)$ and $X \in \mathfrak{diag}(n)$. The role of linearity (vector space) and of commutativity is discussed in Section 4.7. Both $\operatorname{Sym}^{+}(n)$ and $\operatorname{Diag}^{+}(n)$ are convex open subsets of vector spaces.

We continue examples of exponential maps.

(ii) The exponential map gives a surjection(= onto map):

$$\exp : \mathfrak{so}(3) = \{A \mid \text{skew-symmetric}\} \to \{X \mid \text{rotation}\} = SO(3). \tag{4.20}$$

This map is again examined in Section 4.6.

(ii) $\exp : \mathfrak{so}(n) \to SO(n)$ is surjective, but not injective (if $n > 1$).

(i) $\exp : \mathfrak{gl}(n) \to GL^{+}(n)$ is neither surjective nor injective.

(vii) $\exp : \mathfrak{sl}(n) \to SL(n)$ is neither surjective nor injective.

(viii) The exponential map (4.5) to the Lie groups $\mathbb{R}^{n}$ of translations from its Lie algebra gives a bijective map.

(xi) $\exp : \mathfrak{se}(n) \to SE(n)$ is surjective, but not injective.

(x) $\exp : \mathfrak{aff}(n) \to \operatorname{Aff}^{+}(n)$ is neither surjective nor injective.

4.4   ANOTHER DEFINITION OF LIE ALGEBRA

Let G be a matrix group, and $\mathfrak{g} \subset M(n, \mathbb{R})$ its Lie algebra. For $A \in M(n, \mathbb{R})$, the following conditions on φ are equivalent.

- $\varphi(t) = \exp(tA)$.

- $\varphi : \mathbb{R} \to GL(n, \mathbb{R})$ with $\varphi(s + t) = \varphi(s)\varphi(t)$ for all $s, t \in \mathbb{R}$ with $A = \varphi'(0)$.

- $\varphi : \mathbb{R} \to M(n, \mathbb{R})$ with $\varphi(0) = I$ and $\varphi'(t) = A\varphi(t)$.

Such a map φ is called a *one-parameter subgroup* attached to A. Then we know that a one-parameter subgroup takes values in G, i.e., $\varphi : \mathbb{R} \to G$ if and only if $A \in \mathfrak{g}$. This condition characterizes an element A to belong to the Lie algebra $\mathfrak{g}$ in terms of Lie group G.

4.5   LIE ALGEBRA AND DECOMPOSITION

Note that the polar decomposition can be considered as a non-linear and non-commutative counterpart of a linear and commutative natural decomposition, in the level of vector spaces, of square matrices into symmetric and skew-symmetric matrices.

$$\mathfrak{gl}(n) = \mathfrak{so}(n) \oplus \mathfrak{sym}(n). \tag{4.21}$$

We also have such a decomposition as vector spaces

$$\mathfrak{se}(n) = \mathfrak{so}(n) \oplus \mathbb{R}^n, \tag{4.22}$$
$$\mathfrak{aff}(n) = \mathfrak{gl}(n) \oplus \mathbb{R}^n \tag{4.23}$$

for the Lie algebras of semi-direct product groups. Euler angle representation (2.21) can be also regarded as an example: Let

$$J_z = \begin{pmatrix} 0 & -1 & 0 \\ 1 & 0 & 0 \\ 0 & 0 & 0 \end{pmatrix}, \qquad J_y = \begin{pmatrix} 0 & 0 & 1 \\ 0 & 0 & 0 \\ -1 & 0 & 0 \end{pmatrix}, \qquad J_x = \begin{pmatrix} 0 & 0 & 0 \\ 0 & 0 & -1 \\ 0 & 1 & 0 \end{pmatrix}.$$

Then $\{J_z, J_y, J_x\}$ is a basis of $\mathfrak{so}(3)$, so that $\mathfrak{so}(3) = \mathbb{R}J_z \oplus \mathbb{R}J_y \oplus \mathbb{R}J_x$, that is, every skew symmetric matrix of size 3 can be written uniquely as a linear combination of J_x, J_y, and J_z. Note that each $\mathbb{R}J_z, \mathbb{R}J_y, \mathbb{R}J_x$ gives a Lie subalgebra of $\mathfrak{so}(3)$.

In general, let $\mathfrak{g}$ be a Lie algebra, and $X_1, \ldots, X_N \in \mathfrak{g}$ a basis of $\mathfrak{g}$ as a vector space. Then the map from $(t_1, \ldots, t_N)$ to $\exp(t_1 X_1 + t_2 X_2 + \cdots + t_N X_N)$ in G gives a diffeomorphism from a neighborhood at the origin of $\mathbb{R}^N$ to a neighborhood at the identity element of G. This is called *the canonical coordinates of the first kind.* Similarly, the map $\exp(t_1 X_1) \exp(t_2 X_2) \cdots \exp(t_N X_N)$ also gives a local diffeomorphism, which is called *the canonical coordinates of the second kind.* The Euler angle representation is considered to be an example of the canonical coordinates of the second kind. Rodrigues's formula (4.7) is considered to be the canonical coordinates of the first kind. Note that some (multiplicative) matrix decompositions do not give the corresponding (additive) decomposition in a Lie algebra (or a vector space). For example, SVD might give $\mathfrak{gl}(n) \stackrel{?}{=} \mathfrak{so}(n) \oplus \mathfrak{diag}(n) \oplus \mathfrak{so}(n)$, which is obviously wrong because two $\mathfrak{so}(n)$ components have non-trivial intersection. This phenomenon has a relation with the fact that the behavior of SVD at the origin is not stable (not a local diffeomorphism).

4.6 LOSS OF CONTINUITY: SINGULARITIES OF THE EXPONENTIAL MAP

The axis of rotation is a natural invariant of 3D rotation. It is not continuous at the origin of $SO(3)$. This fact has relation with the failure of the local diffeomorphic property of the exponential function. We will explain with some notations: Let $B = \{x \in \mathbb{R}^3 \mid |x| \leq \pi\}$. Then the exponential map restricted to B gives a surjective map

$$\exp : B \to SO(3). \tag{4.24}$$

It is diffeomorphic on the interior

$$\exp : \{x \in \mathbb{R}^3 \mid |x| < \pi\} \to \{R \in SO(3) \mid \det(R + I) \neq 0\}. \tag{4.25}$$

On the boundary, it is a two-to-one covering map

$$\exp : \{x \in \mathbb{R}^3 \mid |x| = \pi\} \to \{R \in SO(3) \mid \det(R + I) = 0\}. \tag{4.26}$$

These two (rather distinct) behaviors are understood in a uniform manner (see Figure 4.2): the map

$$\exp : \{x \in \mathbb{R}^3 \mid 0 < |x| < 2\pi\} \to \{R \in SO(3) \mid R \neq I\} \tag{4.27}$$

gives the two-to-one covering map (everywhere smooth, so that the local inverse does exist uniquely). Slightly more generally, for every integer $n \geq 1$,

$$\exp : \{x \in \mathbb{R}^3 \mid 2(n-1)\pi < |x| < 2n\pi\} \to \{R \in SO(3) \mid R \neq I\} \tag{4.28}$$

also gives the two-to-one covering map. This map factors through the map (2.29):

$$\{x \in \mathbb{R}^3 \mid 2(n-1)\pi < |x| < 2n\pi\} \stackrel{\sim}{\to} \{q \in \mathbb{S}^3 \mid q \neq \pm 1\} \stackrel{2:1}{\to} \{R \in SO(3) \mid R \neq I\}. \tag{4.29}$$

On the other hand, the exponential map on the complement is factored as

$$\{x \in \mathbb{R}^3 \mid |x| = 2n\pi\} \to \{q \in \mathbb{S}^3 \mid q = \pm 1\} \to \{I \in SO(3)\}, \tag{4.30}$$

where the first map is defined as: $x \mapsto (-1)^n$ for $|x| = 2n\pi$. Figure 4.2 illustrates these maps. The first map shows the degeneration of spheres $\{x \in \mathbb{R}^3 \mid |x| = 2n\pi\}$, which looks like circles

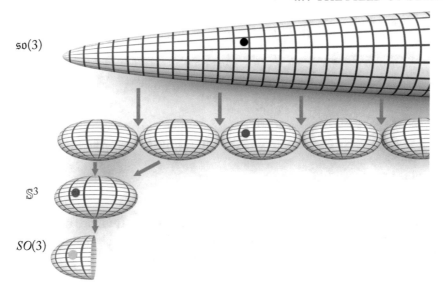

Figure 4.2: The exponential map of $SO(3)$.

in the figure, to a point. By the degeneration (candy-wrapping operation), we obtain $\mathbb{S}^3$ from tube-like body $\{x \in \mathbb{R}^3 \mid 2(n-1)\pi \leq |x| \leq 2n\pi\}$. The second map collects the isomorphic $\mathbb{S}^3$'s for $n = 1, 2, \ldots$ into one piece. The left and right most points in the third stage are 1 and −1 in $\mathbb{S}^3$, which were the joint points on the second stage. The third map is the map (2.29).

   We also understand this phenomenon by the following animation: consider the rotation around x-axis with 360 degrees and after that the rotation around y-axis with certain degree. It seems to be a continuous move, but we do not have a continuous logarithmic lift of this motion. After the first rotation, the transformation (matrix) remembers the axis of rotation, so that the sudden change of the rotation axis from x-axis to y-axis is considered to be a discontinuous move. Note that, if the move does not go through the identity, the continuous logarithmic lift always exists. If we further assume that the move is C^1 (continuously differentiable, that is, the velocity is continuous), then the continuous logarithmic lift exists.

4.7   THE FIELD OF BLENDING

A vector space is, by definition, closed under the interpolation $(1-t)\mathbf{p} + t\,\mathbf{q}$ and the blend $w_1\mathbf{p}_1 + w_2\mathbf{p}_2 + \cdots + w_k\mathbf{p}_k$. In a curved space (such as a group), the interpolation and the blend may not belong to the space again.

   The space where we blend or interpolate something should be a linear space or a convex subset of it (see Figure 4.3). In this sense, the set of rotation matrices is not appropriate, so that it will be replaced by the set of skew symmetric matrices. The set of positive definite symmetric

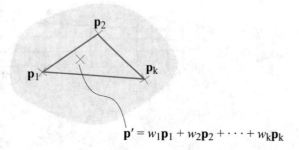

$$\mathbf{p}' = w_1\mathbf{p}_1 + w_2\mathbf{p}_2 + \cdots + w_k\mathbf{p}_k$$

Figure 4.3: Blend in a convex set.

matrices is an open convex subset of the set of symmetric matrices. It sounds not bad for blending, but still the set of symmetric matrices will be better. This is why we once move from one curved space to the other by the exponential function (see Figure 4.4).

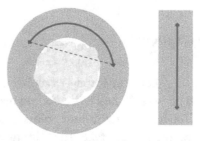

Figure 4.4: Interpolation by linearization.

We also remark that the addition $+$ in the expression $w_1 p_1 + w_2 p_2 + \cdots + w_k p_k$ is a commutative operation: $a + b = b + a$. If the space loses this commutativity, then the interpolation and the blend may not be straightforward. The linearity and commutativity are therefore the key to the interpolation and blending. Note that an interpolation can be considered as a special case of blending; a blend of two things and weights is in between 0 and 1. So the non-commutativity is relaxed for interpolation, but non-linearity still exists.

CHAPTER 5

2D Affine Transformation between Two Triangles

As a simple case, this chapter deals with interpolating two triangles via affine transformations. We note that interpolating affine transformations itself may also have other interesting applications; see, for example, [Shoemake1994b] and [Alexa2002].

5.1 TRIANGLES AND AN AFFINE TRANSFORMATION

The question is to give a continuous interpolation for given two triangles (see Figure 5.1). One way to deal with this question is a linear interpolation of the corresponding vertices. In this chapter, we take an alternative method: use of an interpolation of affine transformation.

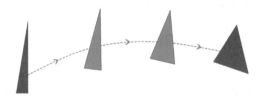

Figure 5.1: Interpolation of triangles.

First of all, we notice that there is a unique affine transformation that maps a given triangle to another one. Specifically, suppose that we are given three points $(x_1, y_1)^T, (x_2, y_2)^T$, and $(x_3, y_3)^T \in \mathbb{R}^2$ forming a triangle, then there is a unique affine transformation

$$\begin{pmatrix} x_1 - x_3 & x_2 - x_3 & x_3 \\ y_1 - y_3 & y_2 - y_3 & y_3 \\ 0 & 0 & 1 \end{pmatrix} = \begin{pmatrix} x_1 & x_2 & x_3 \\ y_1 & y_2 & y_3 \\ 1 & 1 & 1 \end{pmatrix} \begin{pmatrix} 1 & 0 & 0 \\ 0 & 1 & 0 \\ -1 & -1 & 1 \end{pmatrix} \quad (5.1)$$

which maps three points $(1, 0)^T, (0, 1)^T, (0, 0)^T \in \mathbb{R}^2$ into the given three points in this order. In other words, the set of three points forming a triangle is a *principal homogeneous space* of the affine transformation group Aff(2). Suppose that we are given three points $(x_1, y_1)^T, (x_2, y_2)^T$, and $(x_3, y_3)^T \in \mathbb{R}^2$ and want to map them onto other three points $(x'_1, y'_1)^T, (x'_2, y'_2)^T$, and

$(x_3', y_3')^T \in \mathbb{R}^2$ in this order. Then the following 3×3-matrix

$$\hat{A} = \begin{pmatrix} x_1' & x_2' & x_3' \\ y_1' & y_2' & y_3' \\ 1 & 1 & 1 \end{pmatrix} \begin{pmatrix} x_1 & x_2 & x_3 \\ y_1 & y_2 & y_3 \\ 1 & 1 & 1 \end{pmatrix}^{-1} \tag{5.2}$$

is of the form $\begin{pmatrix} a_{1,1} & a_{1,2} & d_x \\ a_{2,1} & a_{2,2} & d_y \\ 0 & 0 & 1 \end{pmatrix}$, and represents the requested affine transformation. We denote

the group of the two-dimensional affine transformations by Aff(2), which are represented by 3×3-matrices of the above form. Note that all the entries a_{ij}, d_x, and d_y are linear in entries of x_j''s. This observation is important in global optimization (see Section 6.3).

We call $A = \begin{pmatrix} a_{1,1} & a_{1,2} \\ a_{2,1} & a_{2,2} \end{pmatrix}$ as the *linear part* and $d_{\hat{A}} = (d_x, d_y)^T$ as the *translation part* of $\hat{A}$
and consider them separately for interpolation. Interpolating the translation part can be neglected (see the discussion in Section 6.3). We focus on interpolation of linear transformation here. In general we may assume that transformation is orientation preserving, that is, it does not flip 2D shapes. We denote the group of the orientation-preserving linear transformations by $GL^+(2)$, which are represented by matrices with positive determinants. We want to interpolate between the identity matrix and a given matrix $A \in GL^+(2)$.

5.2 COMPARISON OF THREE INTERPOLATION METHODS

Next we introduce and compare the following three interpolation methods between the identity matrix and a matrix $A \in GL^+(2)$.

- $A^L(t) := (1-t)I + tA$, linear interpolation.

- $A^P(t) := R_{t\theta} S^L(t) = R_{t\theta}((1-t)I + tS)$ (see [Alexa2000]).

- $A^E(t) := R_\theta^t S^t = R_{t\theta} \exp(t \log S)$ (see [Kaji2012]).

A *homotopy of a linear transformation* $A \in GL^+(2)$ is a series of matrices $A(t)$ parametrized by time $t \in [0, 1]$ such that $A(0) = I$ and $A(1) = A$, where I is the identity matrix. The above three interpolants $A^L(t)$, $A^P(t)$, and $A^E(t)$ satisfy these properties.

The first one A^L gives a linear interpolation in the space of all square matrices $M(n)$. This means that the interpolated matrices can be degenerate (not regular) so that the shape induced by $A^L(t_0)$ might collapse for some t_0.

On the one hand, both homotopies A^P and A^E use the polar decomposition: $A = RS$ [Shoemake1994b]. These homotopies take the same strategy in the sense that they interpolate

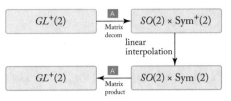

Figure 5.2: ARAP Interpolation in with A^P.

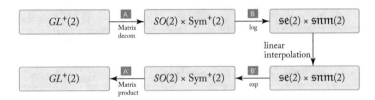

Figure 5.3: Log-Exp interpolation with A^E.

the rotation part and the symmetric part independently, and then get the interpolation of A by multiplying the individual interpolations. Figures 5.2 and 5.3 illustrate A^P and A^E, respectively.

Now we observe more carefully how to interpolate with the homotopies A^P and A^E, referring to Figures 5.4 and 5.5. After the polar decomposition in A^P or A^E, the rotation interpolation at t ($t \in \mathbb{R}$) is the t-th power of matrix R, that is, $R^t = R_{t\theta}$. Note that we can take the angle θ to be $-\pi < \theta \le \pi$, while θ has choice up to modulo 2π. This would cause a problem, which will be discussed in Section 6.3. Through the discussions in Sections 4.1–4.3, we know that this means a simple interpolation technique in Lie group $SO(2)$ through linear interpolation in its Lie algebra $\mathfrak{so}(2)$. First recall that the Lie algebra $\mathfrak{so}(2)$ is the set of matrices in the form of $\begin{pmatrix} 0 & -\theta \\ \theta & 0 \end{pmatrix}$, with $\theta \in \mathbb{R}$. Using the logarithm, we have $\log \begin{pmatrix} \cos\theta & -\sin\theta \\ \sin\theta & \cos\theta \end{pmatrix} = \begin{pmatrix} 0 & -\theta \\ \theta & 0 \end{pmatrix}$. Then it is easy to see that $\mathfrak{so}(2)$ is isomorphic to the real number space $\mathbb{R}$ as a vector space. The rotation interpolation $R_{t\theta}$ can therefore be considered via the exponential map from $\mathfrak{so}(2)$ to $SO(2)$: $\exp \begin{pmatrix} 0 & -t\theta \\ t\theta & 0 \end{pmatrix} = \begin{pmatrix} \cos t\theta & -\sin t\theta \\ \sin t\theta & \cos t\theta \end{pmatrix}$. As for the symmetric factors S, these two interpolations A^P and A^E take the different approaches. A^P interpolates linearly the symmetric S. As illustrated in Figure 5.2, the symmetric part then gives a map from $\text{Sym}^+(2)$ to $\text{Sym}(2)$. This means that the symmetric interpolant in A^P might provide us a degenerate matrix, which is a situation similar to A^L. On the other hand, A^E linearly interpolates $\log(S)$. The "log"operations are done for both rotation and symmetric factors, as illustrated in Figure 5.3. In the special case $A = S \in \text{Sym}^+(2)$, then we have $S^P(t) = (1-t)I + tS$ and $S^E(t) = \exp(t \log(S)) = S^t$. The difference between these two methods is illustrated in Figure 5.6. The method $A^E(t)$ uses both the logarithm and

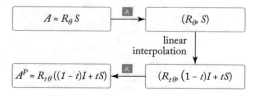

Figure 5.4: Equations in the ARAP interpolation (Figure 5.2).

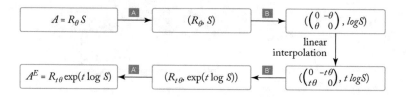

Figure 5.5: Equations in the Log-Exp interpolation (Figure 5.3).

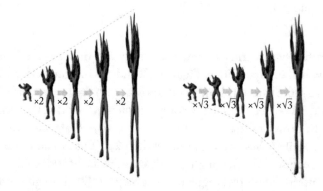

Figure 5.6: Linear interpolation (*left*); exponential interpolation (*right*).

the exponential map, which can be understood a combination of the ideas in [Alexa2002] and [Shoemake1994b]. Throughout this book we refer to this interpolation technique as *Log-Exp interpolation*.

CHAPTER 6

Global 2D Shape Interpolation

6.1 LOCAL TO GLOBAL

Rather than a simple triangle case in Chapter 5, we deal with more general 2D shape interpolation techniques, where we are given two input shapes: source and target (Figure 6.1). We then assume that each shape may be compatibly triangulated.[1] This means that we assume that each shape is triangulated, and that one-to-one correspondence is established between the triangles of the source and target shapes.

Figure 6.1: Problem: Construct a continuous transformation between two triangulated 2D figures.

There are many approaches, including those mentioned earlier, for 2D shape interpolation under the above assumptions. A typical scenario of these approaches came from the seminal work of [Alexa2000]: We first define a homotopy of affine maps for each pair of the corresponding triangles of the source and target objects, such that it connects the identity map and the local affine map that gives a bijection between the corresponding triangles. Let us call this homotopy *local*. Next we construct the homotopy that gives global interpolation between the source and target. This homotopy is defined as a family of the piecewise affine maps, each of which is derived from the affine maps of the local homotopy through a certain energy minimization process. This scenario works well and has inspired many research works.

However, from a practicality viewpoint, there remain many things to be improved and polished. For example, the following practical aspects of the methods should be addressed:

(a) controllability—How to add constraints to get a better result?

(b) rotation consistency—How to treat large rotations (> 180 degrees)?

---

[1]In general, when the two shapes are given without boundary matching nor compatible triangulation, we would need a pre-process to establish them. As for this issue, [Baxter2009] is a good reference describing the most relevant techniques along with their own approach.

(c) symmetry—Can we make it possible that the vertex paths for interpolation from shape A to shape B are the same as from B to A?

Recently [Baxter2008] gave a formulation of rigid shape interpolation using normal equations, presenting the algorithms that meet these requirements.

This section presents a mathematical framework for the above homotopic approaches using affine maps. We start with analyzing the local affine map directly, and introduce a new local homotopy between the affine maps. We also present the algorithms to achieve global interpolation, each of which minimizes an energy function with user-specified constraints. We also discuss how the algorithms meet the above practical requirements. We demonstrate that our mathematical framework gives a comprehensive understanding of rigid interpolation/deformation approaches. In particular we illustrate the power of this framework with the animation examples obtained by several different constraint functions.

6.2   FORMULATION

We now describe the source and target shapes which are compatibly triangulated more explicitly. To make it, we denote the source shape made of triangles by $P = (p_1, \ldots, p_n)$, $(p_i \in \mathbb{R}^2)$, where each p_i is a triangle vertex. Similarly we denote the target shape by $Q = (q_1, \ldots, q_n)$, $(q_i \in \mathbb{R}^2)$, which are the triangle vertices. The triangles are denoted by $\tau_1, \ldots, \tau_m$, where $\tau_i = \{i_1, i_2, i_3\}$ is the set of the indices of the three vertices. Hence, the i-th source (respectively, target) triangle consists of p_{i_1}, p_{i_2}, and p_{i_3} (respectively, q_{i_1}, q_{i_2}, and q_{i_3}) for $\tau_i = \{i_1, i_2, i_3\}$ (Figure 6.2).

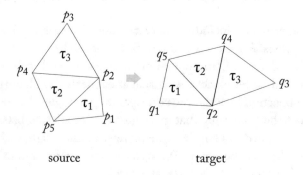

source   target

Figure 6.2: Local affine maps for triangle meshes: $\tau_1 = \{1, 2, 5\}$, $\tau_2 = \{2, 4, 5\}$, $\tau_3 = \{2, 3, 4\}$.

Through Sections 5.1 and 6.3, our local and global interpolation techniques are summarized as follows:

- (from triangle to affine transformation) For each pair of the source and the target triangles corresponding to τ_i, we initially get the affine map $\hat{A}_i \in \text{Aff}(2)$ that maps the initial triangle to the target triangle.

- (local interpolation of linear part) We then construct a homotopy between the 2×2 identity matrix and the linear part A_i of $\hat{A}_i$ (i.e., from $I_2 \in GL^+(2)$ to $A_i \in GL^+(2)$). The homotopy is parametrized by t, with $0 \le t \le 1$. The collection $\{\hat{A}_i \mid i = 1, 2, \ldots, m\}$ of affine maps $\hat{A}_i$'s can be considered as a *piecewise affine transformation* from P to Q (see its precise definition in Section 6.3).

- (global interpolation via error function) We next construct a global homotopy between the inclusion map $P \hookrightarrow \mathbb{R}^2$ and the piecewise affine transformation from P to Q, which will be denoted by $\{\hat{B}_i(t) \in \mathrm{Aff}(2) \mid i = 1, 2, \ldots, m\}$ with $t \in \mathbb{R}$ in Section 6.3. It is obtained by minimizing a global error function regarding the linear part B_i (of $\hat{B}_i$) and A_i along with the user-specified constraint function.

We have been explained the first and the second procedure. We now explain the final procedure.

6.3   ERROR FUNCTION FOR GLOBAL INTERPOLATION

To achieve global interpolation between the two shapes, we have to assemble local transformations considered in Section 5.2 (also see Figure 6.3). In our context, this means that we represent a global

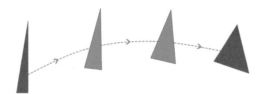

Figure 6.3: Interpolation of triangles.

transformation as a piecewise affine transformation (Figure 6.4). More precisely, we consider a collection of affine maps

$$\mathbf{B}(t) := \{\hat{B}_i(t) \in \mathrm{Aff}(2) \mid i = 1, 2, \ldots, m\}, (0 \le t \le 1)$$

such that $\hat{B}_i(t)$'s are consistent on the edges. More precisely, $\hat{B}_i(t)p_k = \hat{B}_j(t)p_k$ for all t whenever $k \in \tau_i \cap \tau_j$. We put $\mathbf{B}(t)p_k = \hat{B}_i(t)p_k$ for $k \in \tau_i$. Let $v_k(t) := \mathbf{B}(t)p_k, (1 \le k \le n)$ be the image of the initial vertices P. The following observation is vital in this section. The piecewise affine transformation $\mathbf{B}(t)$ which maps p_k's to $v_k(t)$'s is uniquely determined by (5.2) and its entries are linear with respect to $v_k(t)$'s. Therefore, giving $\mathbf{B}(t)$ and giving $v_k(t)$'s are equivalent and we identify them and interchange freely in the following argument. See also Chapter 5. We also assume naturally that

- $\mathbf{B}(t)$ interpolates P and Q, i.e., $v_k(0) = p_k$ and $v_k(1) = q_k$ for all k.

- $B_i(t)$ is "close" to $A_i(t)$, where $B_i(t)$ is the linear part of $\hat{B}_i(t)$ and $A_i(t)$ is the local homotopy obtained in the section.

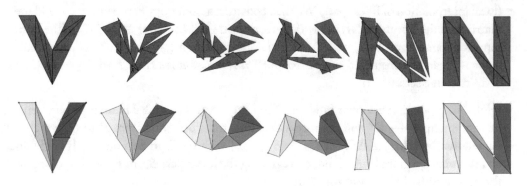

Figure 6.4: From local to global: from one triangle to a triangulated shape.

- Each $B_i(t)$ varies continuously with respect to t.

We will give a framework to obtain global interpolation from given local homotopies. For a moment we consider a fixed t. We then introduce two more ingredients other than local homotopy data;

- a set of *local error functions*

$$E_i : M(2, \mathbb{R}) \times GL^+(2, \mathbb{R}) \to \mathbb{R}_{\geq 0}, \ (1 \leq i \leq m),$$

- a *constraint function*

$$C : (\mathbb{R}^2)^n \to \mathbb{R}_{\geq 0}.$$

The local error function E_i is positive definite and quadratic with respect to the entries of the first factor $M(2, \mathbb{R})$. Intuitively, it measures how different the given two local transformations are. The constraint C is also positive definite and quadratic. It controls the global translation. Furthermore, with this function, we can incorporate various constraints on the vertex path as we will describe later.

If we are given local error functions for each triangle τ_i, $(1 \leq i \leq m)$ and a constraint function, we combine them into a single *global error function*

$$\mathbf{E}_t(\mathbf{B}) := \sum_{i=1}^{m} E_i(B_i(t), A_i(t)) + C(v_1(t), \ldots, v_n(t)),$$

where we regard $\mathbf{B}(t)$ (or more precisely, the entries of $B_i(t)$ which are linear combinations of $v_k(t)$'s as indeterminants to be solved. For each t, the *minimizer* of $\mathbf{E}_t$ may have positive dimension in general, however, one can modify the constraint function C such that it becomes a single point, as we see by concrete examples later. The single minimizer $\mathbf{B}(t)$ is the piecewise affine map that we take as a global interpolation method.

Efficiency of finding the minimizer: We show that finding the minimizer of a global error function is efficient enough. Since the global error function is a positive definite quadratic form, it can be written as a function of $\mathbf{v}(t) = (v_1(t)_x, v_1(t)_y, \ldots, v_n(t)_x, v_n(t)_y)^T \in \mathbb{R}^{2n}$ as

$$E(\mathbf{v}(t)) = \mathbf{v}(t)^T G \mathbf{v}(t) + \mathbf{v}(t)^T \mathbf{u}(t) + c,$$

for some $(2n \times 2n)$-symmetric positive definite matrix G, $\mathbf{u}(t) \in \mathbb{R}^{2n}$, and $c \in \mathbb{R}$. We see that $\mathbf{v}(t) = -\frac{1}{2}G^{-1}\mathbf{u}(t)$ is the minimizer. Note that G is time-independent and we need to compute G^{-1} just once for all frames (see [Alexa2000]).

6.4   EXAMPLES OF LOCAL ERROR FUNCTIONS

In the above point of view, we have a flexibility to choose error functions. For example, we can take

$$E_i^P(B_i(t), A_i(t)) := \sum_{k \in \tau_i} \|B_i(t)p_k - A_i(t)p_k\|^2. \tag{6.1}$$

$$E_i^F(B_i(t), A_i(t)) := \|B_i(t) - A_i(t)\|_F^2, \tag{6.2}$$

$$E_i^S(B_i(t), A_i(t)) := \min_{s, \delta \in \mathbb{R}} \sum_{k \in \tau_i} \|B_i(t)p_k - sR_\delta A_i(t)p_k\|^2, \tag{6.3}$$

$$E_i^R(B_i(t), A_i(t)) := \min_{s, \delta \in \mathbb{R}} \|B_i(t) - sR_\delta A_i(t)\|_F^2, \tag{6.4}$$

where the Frobenius norm of a matrix $M = (m_{ij})$ is defined to be $\|M\|_F^2 = \sum_{i,j} m_{ij}^2$. We now compare these error functions. The error function E^P measures how the intermediate vertices $v_k(t)$'s are different from those obtained by applying the local transformations to the initial vertices. However, this intuitive approach does not produce a good result. We have to speculate on how to define a good error function.

The error function E^F is used in [Alexa2000]. It measures how the local transformation and the final global transformation differ as linear maps. The resulting global error function is invariant under translation and hence requires two-dimensional constraints to get a unique minimizer. For example, [Alexa2000] proposes the following constraint function:

$$C(v_1(t), \ldots, v_n(t)) = \|(1-t)p_1 + tq_1 - v_1(t)\|^2.$$

It produces a fairly satisfactory global transformation when the constraint function is very simple and rotation is "homogeneous." However, this method fails if (a) we want to put some constraints (see Figure 6.5), or (b) the expected rotation angles vary beyond 2π from triangles to triangles (see Figure 6.6).

In order to achieve more flexibility of shape deformation and easier manipulation by a user, [Igarashi2009] and [Igarashi2005] considered error functions which are invariant under similarity

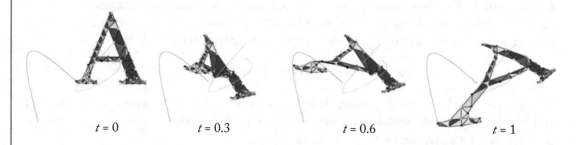

$t = 0$ $t = 0.3$ $t = 0.6$ $t = 1$

Figure 6.5: An example of global interpolation obtained by E_i^F with the constraints on the vertices loci indicated by the curves. In the intermediate frames around $t = 0.3$ and $t = 0.6$, extreme shrink and flip of triangles are observed.

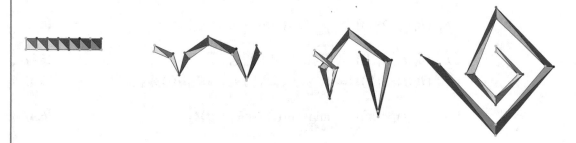

Figure 6.6: An example of global interpolation obtained by E_i^F. To obtain smooth interpolation between the leftmost and rightmost figures, local transformations should deal with rotation angles larger than π, but E_i^F fails to make it.

transformation, i.e., rotation and scale. [Werman1995] has proposed an error function E^S, which is slightly different from them. It measures how different the two sets of points $\{A_i(t)p_k\}$ and $\{B_i(t)p_k\}$ are up to similarity transformation. In [Igarashi2009] and [Igarashi2005] they used a constraint function which forces the vertex loci to be on the specified curves. We will see the detailed construction later.

For the purpose of finding a best-matching global transformation with given local transformations, it is better to use a metric in the space of transformations, rather than in the space of points. The error function E^R, which is a slight modification of E^S, measures how different $A_i(t)$ and $B_i(t)$ are as linear maps up to rotation and scale. The above function E^R has a closed form

$$\min_{s,\delta \in \mathbb{R}} \|sR_\delta A - B\|_F^2 = \|B\|_F^2 - \frac{\|B \cdot A^T\|_F^2 + 2\det(B \cdot A^T)}{\|A\|_F^2}. \tag{6.5}$$

A proof of formula (6.5) is given in Section A.3. This is positive definite quadratic with respect to the entries of B. Since it is invariant under similarity transformation, it avoids the flaws of E_i^F in the cases of (a) and (b); compare Figure 6.7 with Figure 6.5, and Figure 6.8 with Figure 6.6, respectively.

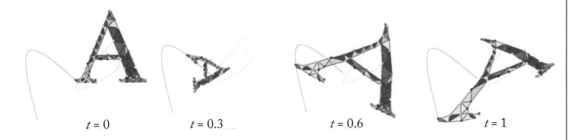

$t = 0$ $t = 0.3$ $t = 0.6$ $t = 1$

Figure 6.7: An example of global interpolation obtained by E_i^R with the same input data as Figure 6.5. By allowing rotational and scale variance without any penalty in the error function, we can get more flexible control of the output animation.

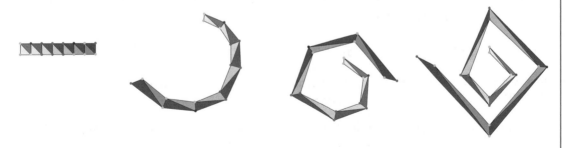

Figure 6.8: An example of global interpolation obtained by E_i^R with the same input data as Figure 6.6. The proper rotation angles for the local triangles are automatically chosen by minimizing the global error function.

We note that a positive linear combination of positive definite quadratic function is also a positive definite quadratic function. This means that a linear combination of above mentioned error functions is also an error function. This idea can be used for practical improvement. We give three examples:

(i) In assembling local error functions, we can take weighted sum instead of ordinary sum. We can put large weights to more important parts (triangles). For example, the more the area of triangle is, the more important its rigidity becomes. Hence, it is reasonable to weight by the areas of the initial triangles:

$$E_i \leftarrow \text{Area}(\Delta(p_{i_1}, p_{i_2}, p_{i_3}))E_i \qquad (\tau_i = \{i_1, i_2, i_3\}).$$

This was already discussed in [Xu2005] and [Baxter2008] as well.

(ii) The local error function E_i^R in (6.4) is employed for a general use. However, we may not want some parts of the 2D shape to rotate or to scale (such as a face of a character). In such cases, we can use a balanced local error function

$$w_i E_i^F(t) + (1 - w_i)E_i^R(t),$$

where $w_i \in [0, 1]$. If we put a large w_i, the rotation and scale of the triangle τ_i would be suppressed. We thus believe that our framework provides more user controllability over previous approaches.

(iii) As is shown in [Baxter2008], we can symmetrize the interpolation by symmetrizing the error function. Let $E_i(t)$ be a global error function for a local homotopies $A_i(t)$, and $E_i^{-1}(t)$ be that for $A_i^{-1}(t)$. Then define a new error function by

$$E_i'(t) := E_i(t) + E_i^{-1}(1 - t).$$

This is symmetric in the sense that it is invariant under the substitution $A_i \leftarrow A_i^{-1}$ and $t \leftarrow 1 - t$. That means that the same minimizing solution is given if we swap the initial and the terminal polygons and reversing time.

6.5   EXAMPLES OF CONSTRAINT FUNCTIONS

Now we give a concise list of the constraints we can incorporate into a constraint function $C(v_1(t), \ldots, v_n(t))$. See the demonstration video in [Kaji2012].

- Some points must trace specified loci (for example, given by B-spline curves). This is realized as follows: let $u_k(t)$ be a user-specified locus of p_k with $u_k(0) = p_k$ and $u_k(1) = q_k$. Then add the term $c_k||v_k(t) - u_k(t)||^2$, where $c_k \geq 0$ is a weight.

- The directions of some edges must be fixed. This is realized by adding the term $c_{kl}||v_k(t) - v_l(t) - e_{kl}(t)||^2$, where $e_{kl}(t) \in \mathbb{R}^2$ is a user-specified vector and $c_{kl} \geq 0$ a weight. This gives a simple way to control the global rotation.

- The barycenter must trace a specified locus $u_o(t)$. This is realized by adding the term $c_o||\frac{1}{n}\sum_{k=1}^{n} v_k(t) - u_o(t)||^2$, where $c_o > 0$ is a weight. This gives a simple way to control the global translation.

Likewise we can add as many constraints as we want.

CHAPTER 7

Parametrizing 3D Positive Affine Transformations

Next we present our 3D application based on the concepts and techniques in Chapters 3 and 4. So let us consider how to parametrize rigid or non-rigid transformations. As we've learned, quaternion or Euler angle parametrizes rotations, and dual quaternion with axis-angle presentation parametrizes the rigid transformation [Kavan2008]. These parametrizations are partial: they deal only with subsets of $Aff^+(3)$, and cannot shear and scale. [Alexa2002] tried to give a Euclidean parametrization of $Aff^+(3)$, considering the *Lie correspondence* between Lie group $Aff^+(3)$ and its Lie algebra through the matrix exponential map and logarithm. However the method is limited for the translations with non-negative eigenvectors, while the Lie correspondence only ensures local bijectivity.

Having in mind these approaches, in this chapter we start with giving an alternative parametrization of $Aff^+(3)$ based on Lie theory [Kaji2013], which successfully parametrizes the whole transformations. Next we show how to integrate the parametrization method to shape deformation techniques as well as Poisson mesh editing [Yu2004]. It will also be demonstrated that these approaches integrated with the parametrization method allow us runtime operations in deforming and animating 3D objects.

7.1 THE PARAMETRIZATION MAP AND ITS INVERSE

Let $M(3, \mathbb{R})$ be the set of 3×3-matrices, as usual. Recall from Section 4.3 that

$$\mathfrak{se}(3) := \left\{ \hat{X} = \begin{pmatrix} X & l \\ 0 & 0 \end{pmatrix} \mid X = -X^T \in M(3, \mathbb{R}), \, l \in \mathbb{R}^3 \right\}$$

is the Lie algebra for the 3-dimensional rigid transformation group $SE(3)$ and that $\mathfrak{sym}(3)$ is the set of 3×3-symmetric matrices. We consider the 12-dimensional parameter space $\mathfrak{se}(3) \times \mathfrak{sym}(3)$ and define the parametrization map (see Figure 7.1)

$$
\begin{aligned}
\phi : \mathfrak{se}(3) \times \mathfrak{sym}(3) &\rightarrow Aff^+(3) \\
\hat{X} \times Y &\mapsto \exp(\hat{X})\iota(\exp(Y)),
\end{aligned}
\tag{7.1}
$$

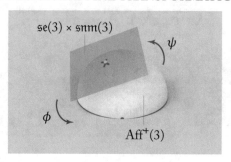

Figure 7.1: The parametrization maps ϕ and ψ.

where exp is the matrix exponential defined by (4.1) in Section 4.1 and $\iota : M(3, \mathbb{R}) \to M(4, \mathbb{R})$ is the natural embedding given by

$$\iota(B) = \begin{pmatrix} B & 0 \\ 0 & 1 \end{pmatrix}.$$

This map ϕ gives a mathematically well-defined parametrization, since it is surjective and has a continuous inverse as we see below. While the above map ϕ is not one-to-one, we can compute its continuous inverse explicitly, thanks to the *Cartan decomposition theorem*. The inverse map ψ is given by

$$\begin{aligned} \psi : \mathrm{Aff}^+(3) &\to \mathfrak{se}(3) \times \mathfrak{sym}(3) \\ \hat{A} &\mapsto \log(\hat{A}\,\iota(\sqrt{A^T A})^{-1}) \times \log(\sqrt{A^T A}). \end{aligned} \qquad (7.2)$$

(See Figure 7.1.) Note that $A^T A$ is a positive definite symmetric matrix so that its square root matrix and its logarithm matrix are uniquely determined. (It is calculated by [Denman1976] and [Cheng2001], for example.) Note also that $\hat{R} := \hat{A}\,\iota(\sqrt{A^T A})^{-1}$ is an element in $SE(3)$ and its logarithm $\log(\hat{R})$ is defined up to modulo 2π.

The computation by the infinite series of the matrix exponential (4.1) is very slow, and hence, it is helpful to give an explicit formula of the matrix exponential and logarithm and a faster algorithm. The explicit form of the matrix exponential and logarithm is given in [Kaji2016] by mimicking the famous Rodrigues's formula [Brockett1984] for 3D rotation matrices. See Section A.2 for the detailed computation of (7.3)–(7.8). Let $\hat{X} \in \mathfrak{se}(3)$ and $\hat{R} \in SE(3)$ with $\exp(\hat{X}) = \hat{R}$ and $\log(\hat{R}) = \hat{X}$. We put $\hat{X} = \begin{pmatrix} X & l \\ 0 & 0 \end{pmatrix}$ and $\hat{R} = \begin{pmatrix} R & d \\ 0 & 1 \end{pmatrix}$. Then the expression of R and d in terms of X and l is given, with an auxiliary parameter θ, by

$$\theta = \sqrt{X_{12}^2 + X_{13}^2 + X_{23}^2} = \sqrt{\frac{1}{2}\mathrm{tr}(X^T X)}, \tag{7.3}$$

$$R = I_3 + \frac{\sin\theta}{\theta}X + \frac{1-\cos\theta}{\theta^2}X^2, \tag{7.4}$$

$$d = \left(I_3 + \frac{1-\cos\theta}{\theta^2}X + \frac{\theta-\sin\theta}{\theta^3}X^2\right)l. \tag{7.5}$$

Conversely, the expression of X and l in terms of R and d is given by

$$\theta = \cos^{-1}\left(\frac{1}{2}\mathrm{tr}(R)\right), \tag{7.6}$$

$$X = \frac{\theta}{2\sin\theta}\left(R - R^T\right), \tag{7.7}$$

$$l = \left(I_3 - \frac{1}{2}X + \frac{2\sin\theta - (1+\cos\theta)\theta}{2\theta^2\sin\theta}X^2\right)d. \tag{7.8}$$

Here we have indeterminacy of $\cos^{-1}$ up to modulo 2π. However, if we impose continuity, we can take one explicit choice. (An explicit code will be given in [Kaji-code].)

We also give a comment on the computation of the matrix exponential and logarithm for the symmetric matrices appearing in (7.1). For any symmetric matrix $Y \in \mathfrak{sym}(3)$, any matrix function can be computed using diagonalization. However, we have introduced [Kaji2016] a faster algorithm to compute the exponential based on the *spectral decomposition* (see [Moler2003], for example). This is because, in most applications, we need to compute (7.2) only once as pre-computation, while computing (7.1) many times in real-time.

7.2   DEFORMER APPLICATIONS

Next we explain the algorithm of our deformers. The input is:

- A target shape to be deformed,

- A set of affine transformations $\{\hat{A}_i \in \mathrm{Aff}(3) \mid 1 \leq i \leq m\}$,

- The weight functions on the vertices $\{w_i : V \to \mathbb{R} \mid 1 \leq i \leq m\}$, where V is the set of the vertices of the target shape.

With the above data, we deform the given shape by the following map:

$$V \ni v \mapsto \phi\left(\sum_{i=1}^{m} w_i \psi(\hat{A}_i)\right)v, \tag{7.9}$$

where we consider the vertex positions $v \in \mathbb{R}^3$ as column vectors and the matrices multiply from the left. Considering the above formulation, we can show the following deformer applications.

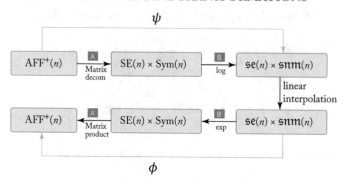

Figure 7.2: Deformation via Log-Exp interpolation.

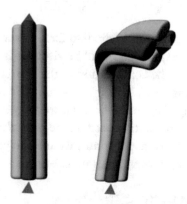

Figure 7.3: The red icons show the probe-based deformer: (*left*) initial positions of the probes; (*right*) the target shape is deformed according to user's manipulation of the probe icons.

Probe-based Deformer

Suppose that a target shape is given. We then assign any number of "probes" which carry transform data. For example, see Figure 7.3, where the red icons mean the probes. If the probes are transformed by the user, the target shape will accordingly be deformed. More precisely, each probe detects the affine map $\hat{A}_i \in \text{Aff}(3)$ which transforms it to the current position from the initial position. A vertex $v \in \mathbb{R}^3$ on the target shape is transformed by Equation (7.9), where the weights w_i's are either painted manually, or computed automatically from the distance between v and the probe location. Figure 7.4 shows another example for designing a vortex shape.

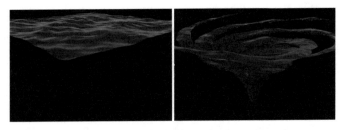

Figure 7.4: Vortex by probe-based deformer: (*left*) initial; (*right*) obtained result.

Cage-based Deformer

Cage-based deformer techniques have been quite popular through recent progress found in [Ju2005]. Suppose that a target shape is given along with a "cage" surrounding it. The cage can be any triangulated polyhedron wrapping the target shape. We want to deform the target shape by directly manipulating it but through the proxy cage (see Figure 7.5). Our parametrization can be used in this framework. A tetrahedron is associated with each face triangle by adding its normal vector (see Figure 7.6). Then each face detects the affine map $\hat{A}_i \in \mathrm{Aff}(3)$ which transforms the

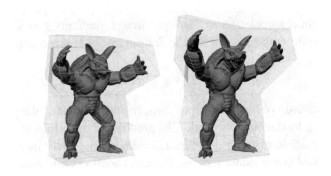

Figure 7.5: Cage-based deformer: (*left*) initialize the cage that surrounding the target; (*right*) the deformed result by user's manipulations on the cage.

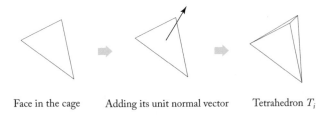

Face in the cage   Adding its unit normal vector   Tetrahedron T_i

Figure 7.6: The key of the algorithm.

initial tetrahedron to the current tetrahedron. A vertex $v \in \mathbb{R}^3$ on the target shape is transformed by Equation (7.9), where the weights w_i's are either painted manually, or computed automatically from the distance between v and the center of the face. For automatic weight computation, it is better to set $w_i = 0$ when v lies in the outer half space of the i-th face.

7.3   INTEGRATING WITH POISSON MESH EDITING

Poisson mesh editing [Yu2004] is a widely used approach to editing triangular mesh by solving the Poisson equation associated. This approach enables users to modify the geometry by editing the boundary conditions for the equation and altering the gradient of mesh implicitly through gradient field manipulation.

In this section we show how the parametrization maps in Section 7.1 are integrated into the framework of Poisson mesh editing. In addition, an improvement of the original method [Yu2004], known as *harmonic guidance* [Zayer2005], will also be discussed and compared with the parametrization map method.

7.3.1   THE POISSON EDITS

The Poisson mesh editing, or more generally *differential surface manipulation* was originally derived from gradient-domain image manipulation. The idea is quite simple (see [Botsch2008], for instance): Under certain manipulation to the input image gradients $\mathbf{g} = \nabla I$, where I denotes either of the pixel intensity (say, of R, G, B), the desired image I' is reconstructed so as to minimize:

$$\int_{\Omega} \|\nabla I' - \mathbf{g}'\|^2 dx\, dy, \qquad (7.10)$$

where Ω denotes the domain of image manipulation, typically a rectangular grid or a region in it, and $\mathbf{g}'$ is the modified $\mathbf{g}$ by the manipulations. The gradient of the resultant image I' is therefore obtained as close as possible to the $\mathbf{g}'$. Variational calculus also gives us an alternative idea to get I'. The I' is obtained as the solution of the Poisson equation with some Direchlet boundary conditions:

$$\Delta I' = \text{div}\, \mathbf{g}'. \qquad (7.11)$$

A typical technique based on this idea is found in [Pérez2003], which also describes the details of the above formulation.

The Poisson Mesh Editing

Roughly speaking, the Poisson mesh editing provides a new mesh geometry under the user-specified constraints (i.e., local edits of the mesh geometry) by solving the Poisson equation, where the constraints are given as the boundary condition associated with the Poisson equation. More specifically, the Poisson equation with Direchlet boundary conditions for mesh editing is formulated as follows:

$$\nabla^2 f = \text{div}\, \mathbf{v}, \quad f|_{\partial\Omega} = f^*|_{\partial\Omega}, \qquad (7.12)$$

where f is an unknown scalar function on the input mesh Ω, f^* means the boundary values as the modeling constraints on the boundary of Ω, and $\mathbf{v}$ is a guidance vector field. The guidance vector field then plays a role similar to the $\mathbf{g}'$ in (7.11) for the case of image-editing. The unknown f actually means one of the three coordinate values of the 3D position of the mesh.

Next we consider numerically solving the Poisson equation for triangle mesh. Then consider f to be the x-coordinate function. In the discretization process the mesh is a piecewise linear surface (i.e., triangle) and then the gradient of the original x-coordinate function is constant on each face. The gradient of the x-coordinate function thus means the projection of the unit x-axis vector $(1, 0, 0)^T$ onto the triangle. The divergence of the guidance vector field in Equation (7.12) can then be expressed as:

$$(\operatorname{div} \mathbf{v})(\mathbf{p}_i) = \sum_{T_k \in N(i)} \nabla B_{ik} \cdot \mathbf{v} |T_k|, \tag{7.13}$$

where $N(i)$ consists of all the triangles connected to the vertex $\mathbf{p}_i$. $|T_k|$ is the area of k-th triangle connected to $\mathbf{p}_i$. B_{ik} is the piecewise linear hat function defined on the vertex. Thus, the divergence defines how much of the guidance vector goes through the mesh at a given vertex. In [Yu2004], it is pointed out that the original formulation based on (7.10) keeps the original gradients $\mathbf{g}'$ constant during deformation. However, the results are not pleasing, because, when we deform a shape, its gradient could usually be changed, i.e., the triangles are transformed during deformation. The most deformed area should then be around the edited area (near the boundary), whereas the other parts far from the boundary should not be affected by the deformation so much.

We therefore need to control the amount of transformation of each triangle. To make it, [Yu2004] introduces the weight for each vertex, which is defined as the distance from the boundary. The weight for each triangle is then defined as the mean of three weight values of its three vertices, which is therefore used to control how much this triangle should be locally transformed. Since the local transform applied to each triangle may be different especially near the edited area, the original mesh will be torn apart and the triangles are not connected any more. We thus need to solve the Poisson equation with the altered guidance vector field, because, as mentioned in [Yu2004], solving the Poisson equation intuitively means stitching together the previously disconnected triangles again. This is the heart of the Poisson mesh edit and would remind us of the deformation techniques described in Chapters 5 and 6. These techniques employ the energy function to be minimized in getting a desired result, while the Poisson mesh edit approaches use the partial differential equation. Also it is noted that the energy function is introduced in solving the global constraints problem, which is specified with the target shape, whereas the Poisson equation deals with the local constraints (user-specified local modifications).

Solving the Poisson equation for mesh editing is thereafter reduced to solving the following linear system for unknown f:

$$Af = \mathbf{b}, \tag{7.14}$$

where A is the discretized differential operator, and **b** is prescribed with the altered guidance vector field based on the boundary conditions and the weights.

The Poisson mesh editing is easy to implement because of its linearity and no requirement to solve any minimization equation, which is normally quite costly in computation. On one hand, the visual quality of deformation by the Poisson mesh editing depends highly on the weights for vertices, even if the boundary conditions are edited properly. In [Yu2004], discrete geodesic distance with some smoothing functions is employed for the weight calculation. Unfortunately it is not intuitive and still suffering from non-smooth weights throughout the mesh.

7.3.2   HARMONIC GUIDANCE

The harmonic field in [Zayer2005] is defined as a scalar field **h** on a mesh satisfying $\nabla^2 \mathbf{h} = 0$. Once the boundary conditions are well defined, the values in the harmonic field are distributed linearly and very smoothly. It is then easy to integrate harmonic field into Poisson mesh editing framework because they share the same Laplacian-operator matrix, A in (7.14), so that re-factorizing the sparse matrix is unnecessary. Once the weights are smoothly distributed over the mesh, the transformation of gradient vectors can be propagated smoothly. In [Zayer2005], quaternion is then used for calculating rotation, while scaling and shear is interpolated linearly. Again this sometimes may cause extreme (or exaggerated) deformations.

7.3.3   THE PARAMETRIZATION MAP FOR POISSON MESH EDITING

We now describe an application of the parametrization map in Section 7.1 to the Poisson mesh editing with the harmonic weights. This means to provide another Poisson mesh editing method using the harmonic weights, which employs the parametrization map for making an arbitrary affine maps available. It is quite natural to integrate the parametrization map with Poisson mesh editing, because our versatile and robust parametrization of affine transformation enables smooth propagation of not only rotation but also scaling and shear from the boundary (user-specified constraints) to interior surface (unknown positions of free vertices on the mesh).

In our implementation, the parametrization map approach is realized as a Maya plug in. Users are therefore able to define an arbitrary number of constraints on the mesh, while editing the shape and creating animation by rotating, translating, and scaling through the handle operations.

In making key-frame animation, these edits are done to make a character's pose at a keyframe, employing pin-and-drag interface as shown in Figure 7.7. Figure 7.8 illustrates that rotation by the parametrization map is much more flexible and easily done, rather than by quaternion in the harmonic guidance approach. Table 7.1 summarizes these methods described above.

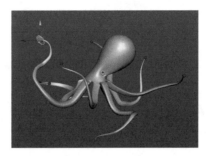

Figure 7.7: The handle interface: The red nodes mean to pin (fix) the edge points of the legs, whereas the red node with coordinate axes, called "handle," shows that the specified vertices are under affine transform operations.

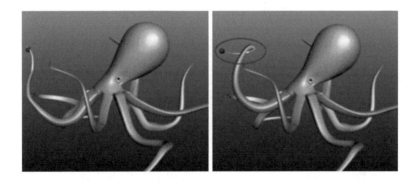

Figure 7.8: Rotation by parametrization map: (*left*) The initial pose of the octopus; (*right*) one of the legs can then be rotated over 360 degrees at runtime operations.

Table 7.1: Poisson mesh editing methods

Method	Weight Distribution	Rotation Representation
Poisson mesh editing	Geodecis distance	Quaternion
Harmonic guidance	Harmonic distance	Quaternion
The parametrization map	Harmonic distance	Log-Exp interpolation

CHAPTER 8

Further Readings

We have presented several mathematical basics that give the theoretical foundation for the deformation/animation techniques in computer graphics. As for your more visual understanding of this book, we would also recommend you to refer to our SIGGRAPH 2016 course notes [Ochiai2016a] and the video assocated with it [Ochiai2016b].

The graphics topics covered in this book are, however, not exhaustive. The mathematics employed in this book then deal only with affine transformations, quaternions, dual quaternions, and their Lie theoretic aspects.

In the following sections, we briefly describe some more mathematical aspects that should further be explored for better understanding of geometric objects and their deformation and animation.

DIFFERENTIAL GEOMETRIC APPROACHES

As shown in Chapters 6 and 7, we need various differential geometric concepts, such as curvature, geodesics, and the fundamental forms, for dealing with more complex structure of deforming geometric objects.

As a good introduction to such differential geometric approaches from the computer scientific viewpoint, we would recommend the following two books: [Kimmel2004] describes differential geometric computational methods and algorithms in image processing and analysis. [Bronstein2009] gives a differential geometric framework for computational models of 2D and 3D non-rigid objects.

Discrete Differential Geometry, DDG, is a rapidly growing new field in mathematics, having natural and fruitful applications in computer graphics. [Hoffmann2009] gives a good introduction to DDG from the mathematician's viewpoint, which we believe gives a nice overview of the theory of discrete curves and surfaces. As for more advanced topics, [Bobenko2008] will give you a nice guidepost, including recent results in DDG.

LIE GROUP APPLICATIONS

Quaternion is a useful tool for controlling rotation. This also suggests that quaternion is powerful in camera control. [Shoemake1994a] proposed a solution of the *camera twist* problem using quaternions, where $\mathbb{S}^3$ is treated as a *fiber bundle*, which means that $\mathbb{S}^3$ is locally homeomorphic to a direct product: (an open subset of 2d sphere)$\times$ (1d circle).

A special class of Lie groups and Lie algebras is realized as (generalized) numbers, such as quaternions. This class of numbers, called geometric algebra or Clifford algebra, has a long history going back to 19th-century mathematician Clifford. A graphical approach is found in [Hanson2006, Dorst2007], and [Dorst2011].

There is another interesting Lie group application, which gives a keyframing method for crowd control [Takahashi2009]. In our context, we regard the method as an interpolation technique in the orthogonal group $O(n)$ of size $n \gg 4$.

Recently the Lie group integrators have been applied to computer animation: [Kobilarov2009] provides a holonomic system for vehicle animations, and [Tournier2012] presents a control system providing trade-off between physics-based simulation and kinematics control using a metric interpolation and real-time animation. The latter one uses PGA, as described next.

Principal geodesic analysis (PGA [Fletcher2004]) provides a new statistical shape analysis method for *data manifold*. The data manifold typically means the collection of shape data, such as of hippocampus in medical imaging [Fletcher2004], or the pose manifold of a motion capture sequence in computer animation [Tournier2012]. As is well known, Principal Component Analysis (PCA) is a standard technique for dimension reduction of statistical data lying on a Euclidian space. PGA is a generalization of this technique for *curved* data manifold. In PGA the concept of Lie group plays a fundamental role and it acts on the data manifold as a symmetric Riemannian manifold.

TOWARD LIE THEORY

How do we describe motion/deformation of objects in $\mathbb{R}^n$? In this book we have shown the following examples:

- Local: Affine transformation well describes a rigid or non-rigid transformation.

- Global: The set of affine maps or Poisson editing approaches well approximate deformation and/or animation.

For global deformation, introducing a diffeomorphism (smooth bijective map whose inverse map is also smooth) may also give us a natural and wider framework of Lie group approaches. We then note that the set of all diffeomorphisms for a given surface (or manifold) constitutes an infinite-dimensional Lie group, where the multiplication is defined as composite of the diffeomorphisms. Unfortunately an infinite-dimensional Lie group is still difficult to understand with the current mathematics, although many attempts have been done. For example, [Mumford2010] gives a good introduction to this general framework, focusing on morphing between two 2D images as input.

Finally, for a more mathematical aspect of Lie groups and Lie algebras, we would like to refer to the classical literatures: [Helgason1978] for Lie groups with differential geometry,

[Knapp1996] for Lie groups with representation theory, [Hochschild1965, Gorbatsevich1993], and [Duistermaat1999] for Lie groups with structure theory. For abstract Lie algebra, we see [Serre1992], and its representation theory with physics application is found in [Georgi1982]. Though those are a bit far from graphics applications, you can consult them to know more about the basic ideas in Lie theory, which we believe will be quite useful for further graphics research.

APPENDIX A

Formula Derivation

In this appendix, we give a few remarks on Rodrigues formulas in Chapters 2 and 4 as well as the energy formula in Chapter 6.

A.1 SEVERAL VERSIONS OF RODRIGUES FORMULA

Rodrigues formulas vary in liturature. Many variation are known and used. For convenience, we summarize the relation between these formula.

$$R(\mathbf{x}) = (\mathbf{u}, \mathbf{x}) + (\cos\theta)\{\mathbf{x} - (\mathbf{u}, \mathbf{x})\mathbf{u}\} + (\sin\theta)(\mathbf{u} \times \mathbf{x}), \tag{A.1}$$
$$R(\mathbf{x}) = (\cos\theta)\mathbf{x} + (1 - \cos\theta)(\mathbf{u}, \mathbf{x})\mathbf{u} + (\sin\theta)(\mathbf{u} \times \mathbf{x}), \tag{A.2}$$
$$R(\mathbf{x}) = \mathbf{x} - (1 - \cos\theta)\{\mathbf{x} - (\mathbf{u}, \mathbf{x})\mathbf{u}\} + (\sin\theta)(\mathbf{u} \times \mathbf{x}), \tag{A.3}$$

$$R = \tag{A.4}$$
$$\begin{pmatrix} \cos\theta + (1 - \cos\theta)u_1^2 & (1 - \cos\theta)u_1u_2 - (\sin\theta)u_3 & (1 - \cos\theta)u_1u_3 + (\sin\theta)u_2 \\ (1 - \cos\theta)u_1u_2 + (\sin\theta)u_3 & \cos\theta + (1 - \cos\theta)u_2^2 & (1 - \cos\theta)u_2u_3 - (\sin\theta)u_1 \\ (1 - \cos\theta)u_1u_3 - (\sin\theta)u_2 & (1 - \cos\theta)u_2u_3 + (\sin\theta)u_1 & \cos\theta + (1 - \cos\theta)u_3^2 \end{pmatrix},$$

$$R = I_3 + (\sin\theta)A + (1 - \cos\theta)A^2, \tag{A.5}$$
$$R = I_3 + \frac{\sin|\mathbf{u}|}{|\mathbf{u}|}A + \frac{1 - \cos|\mathbf{u}|}{|\mathbf{u}|^2}A^2, \tag{A.6}$$
$$R = \exp(\theta A). \tag{A.7}$$

Here $\mathbf{u}$ is assumed to be a unit vector in (A.1), (A.2), (A.3) and (A.4), while $\mathbf{u}$ in (A.6) may not be a unit vector. Note that $\mathbf{u} = (u_1, u_2, u_3)$ in (A.4). Also, for (A.5) and (A.7), $A = -A^T$ should be 'unit', that is, it is assumed that $a_{12}^2 + a_{13}^2 + a_{23}^2 = 1$. In all the cases, $\mathbf{u}$ shows the direction of the rotation axis.

For (A.6), the relation between A and $\mathbf{u}$ is given by $A\mathbf{v} = \mathbf{u} \times \mathbf{v}$. To be more explicit,

$$A = \begin{pmatrix} 0 & -u_3 & u_2 \\ u_3 & 0 & -u_1 \\ -u_2 & u_1 & 0 \end{pmatrix}. \tag{A.8}$$

This means that $|\mathbf{u}| = \sqrt{a_{12}^2 + a_{13}^2 + a_{23}^2}$.

(A.1) explains the meaning as the 2D rotation of angle θ in the plane orthgonal to the vector **u**. In a sense, (A.2) is most popular in the liturature. (A.2) is same as (2.14), and (A.5) is same as (2.15). The direct implications among these equivalent formulas are illustrated as

$$A.1 \leftrightarrow \quad A.2 \quad \leftrightarrow A.4 \leftrightarrow \quad A.5 \quad \leftrightarrow A.7$$
$$\updownarrow \qquad\qquad\qquad \updownarrow$$
$$A.3 \qquad\qquad\qquad A.6$$

A.2   RODRIGUES TYPE FORMULA FOR MOTION GROUP

We explain the computation of formulas (7.3)–(7.8), where (7.5) and (7.8) might be less well-known, as compared with the others.

Having $\hat{X} \in \mathfrak{se}(3)$ in Section 7.1, we notice that

$$\hat{X}^k = \begin{pmatrix} X^k & kX^{k-1}l \\ 0 & 0 \end{pmatrix} \qquad \text{for } k = 1, 2, \dots.$$

Then

$$\exp(\hat{X}) = \sum_{k=0}^{\infty} \frac{1}{k!} \begin{pmatrix} X^k & kX^{k-1}l \\ 0 & 0 \end{pmatrix} = \begin{pmatrix} \exp(X) & Yl \\ 0 & 1 \end{pmatrix},$$

where we define

$$Y = \sum_{k=1}^{\infty} \frac{1}{k!} X^{k-1}. \tag{A.9}$$

This shows $R = \exp(X)$ and $d = Yl$.

The relations (7.3),(7.4),(7.6) and (7.7) on X and R are known as Rodrigues formulas. Actually, (7.4) is essentially (A.5). (7.3) coincides with the requirement on the relation between A and **u** in (A.5).

(7.3) gives

$$2\theta^2 = -\text{tr}(X^2).$$

Taking the trace of (7.4), together with $\text{tr}(X) = 0$, we obtain

$$\text{tr}(R) = 3 + \frac{1 - \cos\theta}{\theta^2}\text{tr}(X^2) = 3 - 2(1 - \cos\theta) = 1 + 2\cos\theta,$$

which is equivalent to (7.6).

By (7.4), we obtain

$$R - R^T = \frac{2\sin\theta}{\theta}X,$$

which is equivalent to (7.7).

Finally, we come to (7.5) and (7.8). What we should do is to derive

$$Y = I_3 + \frac{1 - \cos \theta}{\theta^2} X + \frac{\theta - \sin \theta}{\theta^3} X^2, \tag{A.10}$$

$$Y^{-1} = I_3 - \frac{1}{2} X + \frac{2 \sin \theta - (1 + \cos \theta)\theta}{2\theta^2 \sin \theta} X^2. \tag{A.11}$$

We will obtain these formulas by noticing $(X^2 + \theta^2 I_3)X = O$. For (A.10),

$$Y = I + \sum_{m=1}^{\infty} \frac{1}{(2m)!} X^{2m-1} + \sum_{m=1}^{\infty} \frac{1}{(2m+1)!} X^{2m} +$$

$$= I + \sum_{m=1}^{\infty} \frac{1}{(2m)!} (-\theta^2)^{m-1} X + \sum_{m=1}^{\infty} \frac{1}{(2m+1)!} (-\theta^2)^{m-1} X^2$$

$$= I + \frac{1 - \cos \theta}{\theta^2} X + \frac{\theta - \sin \theta}{\theta^3} X^2.$$

An alternative explanation of (A.10) will be the following. We assume that Y is of the form $Y = I_3 + a_1 X + a_2 X^2$. We compute XY in two ways:

$$XY = X + a_1 X^2 + a_2 X^3 = (1 - a_2 \theta^2)X + a_1 X^2,$$

$$XY = \sum_{k=1}^{\infty} \frac{1}{k!} X^k = \exp(X) - I = \frac{\sin \theta}{\theta} X + \frac{1 - \cos \theta}{\theta^2} X^2.$$

By comparing the coefficients, we obtain $1 - a_2 \theta^2 = (\sin \theta)/\theta$ and $a_1 = (1 - \cos \theta)/\theta^2$, which proves (A.10).

For (A.11), we put $Y = I_3 + a_1 X + a_2 X^2$ and $Y^{-1} = I_3 + b_1 X + b_2 X^2$, and write the equation

$$\begin{aligned} I_3 = YY^{-1} &= (I_3 + a_1 X + a_2 X^2)(I_3 + b_1 X + b_2 X^2) \\ &= I_3 + (a_1 + b_1)X + (a_2 + b_2 + a_1 b_1)X^2 + (a_1 b_2 + a_2 b_1)X^3 + a_2 b_2 X^4 \\ &= I_3 + (a_1 + b_1)X + (a_2 + b_2 + a_1 b_1)X^2 - (a_1 b_2 + a_2 b_1)\theta^2 X - a_2 b_2 \theta^2 X^2. \end{aligned}$$

Suppose $a_1 + b_1 - (a_1 b_2 + a_2 b_1)\theta^2 = 0$ and $(a_2 + b_2 + a_1 b_1) - a_2 b_2 \theta^2 = 0$. This requirement is reduced to the system of linear equations in unknown variables b_1 and b_2 as

$$\begin{pmatrix} a_2 \theta^2 - 1 & a_1 \theta^2 \\ -a_1 & a_2 \theta^2 - 1 \end{pmatrix} \begin{pmatrix} b_1 \\ b_2 \end{pmatrix} = \begin{pmatrix} a_1 \\ a_2 \end{pmatrix},$$

which can be solved as

$$\begin{pmatrix} b_1 \\ b_2 \end{pmatrix} = \begin{pmatrix} a_2 \theta^2 - 1 & a_1 \theta^2 \\ -a_1 & a_2 \theta^2 - 1 \end{pmatrix}^{-1} \begin{pmatrix} a_1 \\ a_2 \end{pmatrix} = \frac{1}{a_1^2 \theta^2 + (1 - a_2 \theta^2)^2} \begin{pmatrix} -a_1 \\ a_1^2 - a_2(1 - a_2 \theta^2) \end{pmatrix}.$$

If we put the explicit value of a_1 and a_2, then we obtain

$$\begin{pmatrix} b_1 \\ b_2 \end{pmatrix} = \begin{pmatrix} -1/2 \\ \frac{2\sin\theta - (1+\cos\theta)\theta}{2\theta^2\sin\theta} \end{pmatrix}.$$

This is the required formula (A.11) for the inverse of Y.

A.3   PROOF OF THE ENERGY FORMULA

We give a proof of the formula (6.5)

$$\min_{s,\delta\in\mathbb{R}} \|sR_\delta A - B\|_F^2 = \|B\|_F^2 - \frac{\|B \cdot A^T\|_F^2 + 2\det(B \cdot A^T)}{\|A\|_F^2}.$$

First note that the set of all matrices of the form sR_δ is a vector space of skew-symmetric matrices:

$$\{sR_\delta \mid s,\delta \in \mathbb{R}\} = \{xI + yJ \mid x, y \in \mathbb{R}\}.$$

Here we denote

$$I = \begin{pmatrix} 1 & 0 \\ 0 & 1 \end{pmatrix}, \quad J = \begin{pmatrix} 0 & -1 \\ 1 & 0 \end{pmatrix}.$$

So the left-hand side of the problem is rewritten as

$$\min_{s,\delta\in\mathbb{R}} \|sR_\delta A - B\|_F^2 = \min_{x,y\in\mathbb{R}} \|(xI + yJ)A - B\|_F^2.$$

Recall the definition of Frobenius norm:

$$\|A\|_F^2 = \operatorname{tr}(AA^T),$$

where tr denotes the trace of a square matrix. Then

$$\begin{aligned}
&\|(xI + yJ)A - B\|_F^2 \\
&= \operatorname{tr}((xA + yJA - B)(xA^T + yA^TJ^T - B^T)) \\
&= \operatorname{tr}(x^2AA^T + xyAA^TJ + xyJAA^T + y^2JAA^TJ^T \\
&\qquad - xAB^T - xBA^T - yJAB^T - yBA^TJ^T + BB^T) \\
&\overset{(*)}{=} \operatorname{tr}(x^2AA^T + y^2JAA^TJ^T - 2xBA^T - 2yBA^TJ^T + BB^T) \\
&= x^2\|A\|_F^2 + y^2\|JA\|_F^2 - 2x\operatorname{tr}(BA^T) - 2y\operatorname{tr}(BA^TJ^T) + \|B\|_F^2.
\end{aligned}$$

Here, for the computation (*), we have used

$$\begin{aligned}
S^T &= S \text{ if } S = AA^T, & \text{(A.12)} \\
\operatorname{tr}(SJ) &= 0 \text{ if } S^T = S, & \text{(A.13)} \\
\operatorname{tr}(PQ^T) &= \operatorname{tr}(QP^T). & \text{(A.14)}
\end{aligned}$$

Now we will use

$$\|JA\|_F^2 = \operatorname{tr}(JAA^T J^T) = \operatorname{tr}(AA^T J^T J) = \operatorname{tr}(AA^T) = \|A\|_F^2,$$

since $J^T J = -J^2 = I$. This shows

$$
\begin{aligned}
&\|(xI + yJ)A - B\|_F^2 \\
&= x^2\|A\|_F^2 + y^2\|A\|_F^2 - 2x\operatorname{tr}(BA^T) - 2y\operatorname{tr}(BA^T J^T) + \|B\|_F^2 \\
&= \|A\|_F^2\left(x^2 - 2x\frac{\operatorname{tr}(BA^T)}{\|A\|_F^2}\right) + \|A\|_F^2\left(y^2 - 2y\frac{\operatorname{tr}(BA^T J^T)}{\|A\|_F^2}\right) + \|B\|_F^2 \\
&= \|A\|_F^2\left(x - \frac{\operatorname{tr}(BA^T)}{\|A\|_F^2}\right)^2 - \frac{(\operatorname{tr}(BA^T))^2}{\|A\|_F^2} + \|A\|_F^2\left(y - \frac{\operatorname{tr}(BA^T J^T)}{\|A\|_F^2}\right)^2 \\
&\quad - \frac{(\operatorname{tr}(BA^T J^T))^2}{\|A\|_F^2} + \|B\|_F^2 \\
&= \|B\|_F^2 - \frac{\|B \cdot A^T\|_F^2 + 2\det(B \cdot A^T)}{\|A\|_F^2} \\
&\quad + \|A\|_F^2\left(x - \frac{\operatorname{tr}(BA^T)}{\|A\|_F^2}\right)^2 + \|A\|_F^2\left(y - \frac{\operatorname{tr}(BA^T J^T)}{\|A\|_F^2}\right)^2.
\end{aligned}
$$

Here in the last equality, we have used the following identity

$$(\operatorname{tr}Q)^2 + (\operatorname{tr}(QJ^T))^2 = \|Q\|_F^2 + 2\det(Q).$$

This identity is equivalent, if we put $Q = \begin{pmatrix} a & b \\ c & d \end{pmatrix}$, to the following identity

$$(a + d)^2 + (-b + c)^2 = (a^2 + b^2 + c^2 + d^2) + 2(ad - bc),$$

which will be examined by expanding the left-hand side.

Now we see that the minimum is taken at

$$x = \frac{\operatorname{tr}(BA^T)}{\|A\|_F^2}, \quad y = \frac{\operatorname{tr}(BA^T J^T)}{\|A\|_F^2}$$

and its minimum value is given by

$$\|B\|_F^2 - \frac{\|B \cdot A^T\|_F^2 + 2\det(B \cdot A^T)}{\|A\|_F^2},$$

which is the desired result. This is the end of the proof.

Bibliography

[Alexa2000] M. Alexa, D. Cohen-Or, and D. Levin, As-rigid-as-possible shape interpolation, *SIGGRAPH Proc. of the 27th Annual Conference on Computer Graphics and Interactive Techniques*, pages 157–164, 2000. DOI: 10.1145/344779.344859. 42, 45, 49

[Alexa2002] M. Alexa, Linear combinations of transformations, In *ACM Transactions on Graphics (TOG)—Proc. of ACM SIGGRAPH*, 21(3), pages 380–387, 2002. DOI: 10.1145/566654.566592. 29, 41, 44, 53

[Baxter2008] W. Baxter, P. Barla, and K. Anjyo, Rigid shape interpolation using normal equations, In *NPAR Proc. of the 6th International Symposium on Non-photorealistic Animation and Rendering*, pages 59–64, 2008. DOI: 10.1145/1377980.1377993. 46, 52

[Baxter2009] W. Baxter, P. Barla, and K. Anjyo, Compatible embedding for 2D shape animation, *IEEE Transactions on Visualization and Computer Graphics*, 15(5), pages 867–879, 2009. DOI: 10.1109/TVCG.2009.38. 45

[Bobenko2008] A.I. Bobenko, P. Schröder, J.M. Sullivan, and G.M. Ziegler (Eds.), *Discrete Differential Geometry*, Obervolfach Seminars vol. 38, Birkhäuser, 2008. DOI: 10.1007/978-3-7643-8621-4. 63

[Botsch2008] M. Botsch and O. Sorkine, On linear variational surface deformation methods, *IEEE Transactions on Visualization and Computer Graphics*, 14(1), pages 213–230, 2008. DOI: 10.1109/TVCG.2007.1054. 58

[Brockett1984] R.W. Brockett, Robotic manipulators and the product of exponentials formula, *Mathematical Theory of Networks and Systems*, Lecture Notes in Control and Information Sciences, 58, pages 120–129, 1984. DOI: 10.1007/BFb0031048. 54

[Bronstein2009] A.M. Bronstein, M.M. Bronstein, and R. Kimmel, *Numerical Geometry of Non-Rigid Shapes*, Springer-Verlag, 2009. 63

[Chaudhry2010] E. Chaudhry, L.H. You, and J.J. Zhang, Character skin deformation: A survey, *Proc. of the 7th International Conference on Computer Graphics, Imaging and Visualization (CGIV2010)*, pages 41–48, IEEE, 2010. DOI: 10.1109/CGIV.2010.14. 29

[Cheng2001] S.H. Cheng, N.J. Higham, C.S. Kenney, and A.J. Laub, Approximating the logarithm of a matrix to specified accuracy, *SIAM Journal on Matrix Analysis and Applications*, 22(4), pages 1112–1125, 2001. DOI: 10.1137/S0895479899364015. 54

[Denman1976] E.D. Denman and A.N. Beavers, *The matrix sign function and computa-tions in systems, Applied Mathematics and Computation*, 2(1), pages 63–94, 1976. DOI: 10.1016/0096-3003(76)90020-5. 54

[Dorst2007] L. Dorst, D. Fontijne, and S. Mann, *Geometric Algebra for Computer Science*, Morgan and Kaufmann, 2007. 64

[Dorst2011] L. Dorst and J. Lasenby, (Ed.), *Guide to Geometric Algebra in Practice*, Springer-Verlag, 2011. DOI: 10.1007/978-0-85729-811-9. 64

[Duistermaat1999] J.J. Duistermaat and J.A.C. Kolk, *Lie Groups*, Springer, Universitext, 1999. 65

[Ebbinghaus1991] H.-D. Ebbinghaus, H. Hermes, F. Hirzebruch, M. Koecher, K. Mainzer, J. Neukirch, A. Prestel, and R. Remmert, *Numbers*, Graduate Texts in Mathematics, Springer-Verlag, 1991. DOI: 10.1007/978-1-4612-1005-4. 3, 13

[Fletcher2004] P.T. Fletcher, C.Lu, S.M. Pizer, and S. Joshi, Principal geodesic analysis for the study of nonlinear statistics of shape, *IEEE Transactions on Medical Imaging*, 23(8), pages 995-1005, 2004. DOI: 10.1109/TMI.2004.831793. 64

[Georgi1982] H. Georgi, *Lie Algebras in Particle Physics, from Isospin to Unified Theories*, Ben-jamin/Cummings Publishing Co., Inc., 1982. 65

[Gorbatsevich1993] V.V. Gorbatsevich, A.L. Onishchik, and E.B. Vinberg, *Foundations of Lie Theory and Lie Transformation Groups*, Springer-Verlag, 1993. Originally published as Lie Groups and Lie Algebras I, vol. 20 of the Encyclopaedia of Math. Sci., Springer-Verlag. 65

[Hanson2006] A. Hanson, *Visualizing Quaternions*, Morgan-Kaufmann/Elsevier, 2006. 13, 64

[Helgason1978] S. Helgason, *Differential Geometry, Lie Groups, and Symmetric Spaces*, Academic Press, 1978, reprinted by the American Mathematical Society, 2001. 28, 64

[Hochschild1965] G. Hochschild, *The Structure of Lie Groups*, Holden-Day, Inc., Amsterdam, 1965. 65

[Hoffmann2009] T. Hoffmann, *Discrete Differential Geometry of Curves and Surfaces*, MI Lecture Note Series, Kyushu University, vol. 16, 2009. 63

[Igarashi2005] T. Igarashi, T. Moscivich, and J.F. Hughes, As-rigid-as-possible shape manipula-tion, *ACM Transactions on Graphics (TOG)—Proc. of ACM SIGGRAPH*, 24(3), pages 1134–1141, 2005. 49, 50

[Igarashi2009] T. Igarashi and Y. Igarashi, Implementing as-rigid-as-possible shape manipula-tion and surface flattening, *Journal of Graphics, GPU, and Game Tools*, 14(1), pages 17–30, 2009. 49, 50

[Ju2005] T. Ju, S. Schaefer, and J. Warren, Mean value coordinates for closed triangular meshes, *ACM Transactions on Graphics (TOG)—Proc. of ACM SIGGRAPH* 24(3), pages 561–566, 2005. DOI: 10.1145/1073204.1073229. 57

[Kaji2012] S. Kaji, S. Hirose, S. Sakata, Y. Mizoguchi, and K. Anjyo, Mathematical analysis on affine maps for 2D shape interpolation, *SCA Proc. of the ACM SIGGRAPH/Eurographics Symposium on Computer Animation* , pages 71–76, 2012 DOI: 10.2312/SCA/SCA12/071-076. 42, 52

[Kaji2013] S. Kaji, S. Hirose, H. Ochiai, and K. Anjyo, A lie theoretic parameterization of affine transformations, *Proc. MEIS2013 Symposium: Mathematical Progress in Expressive Image Synthesis*, MI Lecture Note Series vol. 50, pages 134–140, 2013. DOI: 10.1145/2542266.2542268. 53

[Kaji2016] S. Kaji and H. Ochiai, A concise parametrization of affine transformation, *SIAM J. Imaging Sci.*, 9(3), pages 1355–1373, 2016. 10.1137/16M1056936. 54, 55

[Kaji-code] S. Kaji, A c++ library for 3d affine transformation. https://github.com/shizuo-kaji/AffineLib, 2014. 55

[Kavan2008] L. Kavan, S. Collins, J. Zara, and C. O'Sullivan, Geometric skinning with approximate dual quaternion blending, *ACM Transactions on Graphics (TOG)*, 27(4), Article 105, 2008. DOI: 10.1145/1409625.1409627. 17, 36, 53

[Kimmel2004] R. Kimmel, *Numerical Geometry of Images*, Springer-Verlag, 2004. DOI: 10.1007/978-0-387-21637-9. 63

[Klein1926] F. Klein, *Vorlesungen über die Entwicklung der Mathematik im 19.* Jahrhundert I, Springer-Verlag, 1926. See also, *Development of Mathematics in the 19th century*, translated by M. Ackerman, Math. Sci. Press, 1979. 25

[Knapp1996] A. Knapp, *Lie Groups, Beyond an Introduction*, Birkhäuser, 1996. DOI: 10.1007/978-1-4757-2453-0. 65

[Kobilarov2009] M. Kobilarov, K. Crane, and M. Desbrun, Lie group integrators for animation and control of vehicles, *ACM Transactions on Graphics (TOG)*, 28(6), Article 16, 2009. DOI: 10.1145/1516522.1516527. 64

[Lewis2000] J.P. Lewis, M. Cordner, and N. Fong, Pose space deformation: A unified approach to shape interpolation and skeleton-driven deformation, *SIGGRAPH Proc. of the 27th Annual Conference on Computer Graphics and Interactive Techniques*, pages 165–172, 2000. DOI: 10.1145/344779.344862. 29

[Matsuda2004] G. Matsuda, S. Kaji, and H. Ochiai, Anti-commutative dual 2d rigid transformation, *Mathematical Progress in Expressive Image Synthesis I*, Springer-Verlag, 2014. DOI: 10.1007/978-4-431-55007-5_17. 19

[Moler2003] C. Moler and C. van Loan, Nineteen dubious ways to compute the exponential of a matrix, twenty-five years later, *SIAM Review*, 45(1), pages 3–49, 2003. DOI: 10.1137/S00361445024180. 55

[Mumford2010] D. Mumford and A. Desoineux, *Pattern Theory*, A.K. Peters, 2010. 64

[Nieto2013] J.R. Nieto and A. Susín, Cage based deformations: A survey, *Deformation Models*, M.G. Hidalgo, A.M. Torres and Javier Varona Gómez (Eds.) Lecture Notes in Computational Vision and Biomechanics 7, 2013. DOI: 10.1007/978-94-007-5446-1_3. 29

[Ochiai2016a] H. Ochiai, K. Anjyo and A. Kimura, An elementary introduction to matrix exponential for CG, *SIGGRAPH Courses*, Article No. 4, 2016. DOI: 10.1145/2897826.2927338. 63

[Ochiai2016b] H. Ochiai, K. Anjyo and A. Kimura, Mathematical basics for computer graphics, 2016. https://youtu.be/I2Y-pJYmu9A. 63

[Pérez2003] P. Pérez, M. Gangnet, and A. Blake, Poisson image editing, *ACM Transactions on Graphics (TOG)*, 22(3), pages 313–318, 2003. DOI: 10.1145/882262.882269. 58

[Serre1992] Jean-Pierre Serre, Lie algebras and lie groups. 1964 lectures given at Harvard University, 2nd ed., *Lecture Notes in Mathematics*, 1500, Springer-Verlag, Berlin, 1992. 35, 65

[Shoemake1985] K. Shoemake, Animating rotation with quaternion curves, In *SIGGRAPH Proc. of the 12th Annual Conference on Computer Graphics and Interactive Techniques*, pages 245–254, 1985. DOI: 10.1145/325334.325242. 13

[Shoemake1994a] K. Shoemake, Fiber bundle twist reduction, *Graphics Gems IV*, Academic Press, pages 230–236, 1994. DOI: 10.1016/B978-0-12-336156-1.50031-8. 63

[Shoemake1994b] K. Shoemake, Quaternions, 1994. http://www.cs.ucr.edu/~vbz/resources/quatut.pdf 41, 42, 44

[Stubhaug2002] A. Stubhaug, *The Mathematician Sophus Lie—It was the Audacity of My Thinking*, Springer-Verlag, 2002. The original publication in Norwegian is in 2000 from H. Aschehoug & Co. DOI: 10.1007/978-3-662-04386-8. 33

[Takahashi2009] S. Takahashi, K. Yoshida, T. Kwon, K.H. Lee, J. Lee, and S.Y. Shin, Spectral-based group formation control, *Computer Graphics Forum*, 28(2), pages 639–648. 2009. DOI: 10.1111/j.1467-8659.2009.01404.x. 64

[Tournier2009] M. Tournier, X. Wu, N. Courty, E. Amaud, and L. Revéret, Motion compression using principal geodesics analysis, *Computer Graphics Forum*, 28(2), pages 355–364, 2009. DOI: 10.1111/j.1467-8659.2009.01375.x. 29

[Tournier2012] M. Tournier and L. Revéret, Principal geodesic dynamics, *SCA Proc. of the ACM SIGGRAPH/Eurographics Symposium on Computer Animation*, pages 235–244, 2012. DOI: 10.2312/SCA/SCA12/235-244. 64

[Vince2011] J. Vince, *Quaternions for Computer Graphics*, Springer-Verlag, 2011. DOI: 10.1007/978-0-85729-760-0. 13

[Watt1992] A. Watt and M. Watt, *Advanced Animation and Rendering Techniques*, Addison Wesley, 1992. 13

[Werman1995] M. Werman and D. Weishall, Similarity and affine invariant distances between 2d point sets, *IEEE Transactions on Pattern Analysis and Machine Intelligence*, 17(18), pages 810–814, 1995. DOI: 10.1109/34.400572. 50

[Xu2005] D. Xu, H. Zhang, Q. Wang, and H. Bao, Poisson shape interpolation, *SPM Proc. of the ACM Symposium on Solid and Physical Modeling*, pages 267–274, 2005. DOI: 10.1145/1060244.1060274. 52

[Yu2004] Y. Yu, K. Zhou, D. Xu, X. Shi, H. Bao, B. Guo, and H.-Y. Shum, Mesh editing with poisson-based gradient field manipulation, *ACM Transactions on Graphics (TOG)—Proc. of ACM SIGGRAPH*, 23(3), pages 644–651, 2004. DOI: 10.1145/1015706.1015774. 53, 58, 59, 60

[Zayer2005] R. Zayer, C. Rssl, Z. Karni, and H.-P. Seidel, Harmonic guidance for surface deformation, *Computer Graphics Forum*, 24(3), pages 601–609, 2005. DOI: 10.1111/j.1467-8659.2005.00885.x. 58, 60

Authors' Biographies

KEN ANJYO

Ken Anjyo is the R&D supervisor at OLM Digital. He has been credited as R&D supervisor for recent Pokémon and several other movies. His research focuses on construction of mathematical and computationally tractable models. Dr. Anjyo's research includes recent SIGGRAPH and IEEE CG&A papers on art-directable specular highlights and shadows for anime, the Fourier method for editing motion capture, and direct manipulation blendshapes for facial animation. He is co-founder of the Digital Production Symposium (DigiPro) that started in 2012 and served as SIGGRAPH Asia 2015 Course co-chair, SIGGRAPH 2014 and 2015 Computer Animation Festival juror, and co-founder of the Mathematical Progress in Expressive Image Synthesis symposium (MEIS). He is appointed as the SIGGRAPH Asia 2018 conference chair. He is also a VES member since 2011. http://anjyo.org.

HIROYUKI OCHIAI

Hiroyuki Ochiai is a Professor at Institute of Mathematics for Industry, Kyushu University, Japan. He received his Ph.D. in mathematics from the University of Tokyo in 1993. His research interests include representation theory of Lie groups and Lie algebras, algebraic analysis, and group theory. He has been joining the CREST project *Mathematics for Computer Graphics* led by Ken Anjyo since 2010. He was a lecturer of courses at SIGGRAPH Asia 2013, SIGGRAPH2014 and 2016: http://mcg.imi.kyushu-u.ac.jp/.